D1557546

Nutrient Elements and Toxicants

Comparative Animal Nutrition

Vol. 2

Series Editor: M. RECHCIGL, Jr., Washington, D. C.

S. Karger · Basel · München · Paris · London · New York · Sydney

Nutrient Elements and Toxicants

Editor: M. Rechcigl, Jr., Washington, D. C.

10 figures and 19 tables, 1977

S. Karger · Basel · München · Paris · London · New York · Sydney

Comparative Animal Nutrition

Vol. 1: Carbohydrates, Lipids and Accessory Growth Factors
Editor: M. RECHCIGL, Jr., Washington, D.C.
XII + 224 p., 30 fig., 25 tab., 1976
ISBN 3-8055-2268-1

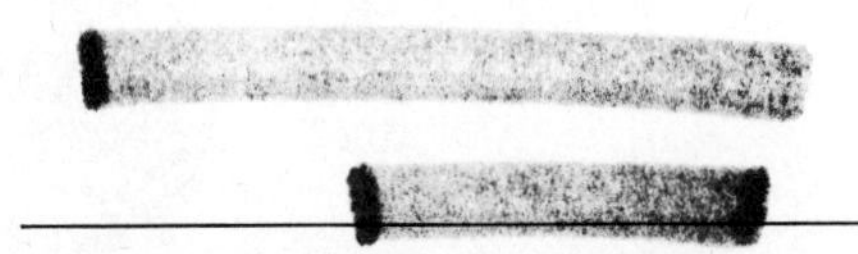

Cataloging in Publication
Nutrient elements and toxicants
Editor: M. Rechcigl, Jr. – Basel, New York: Karger, 1977
(Comparative animal nutrition; v. 2)
1. Animal Nutrition 2. Food Poisoning – veterinary
I. Rechcigl, Miloslav, 1930–, ed. II. Title III. Series
W1 CO434N v. 2/SF 95 N976
ISBN 3-8055-2351-3

Printed in Switzerland by Meier + Cie AG Schaffhausen
ISBN 3-8055-2351-3

Contents

Preface IX
Abbreviations and Symbols XI

The Chemical Element Needs of Animals
ERNEST BEERSTECHER

I. Introduction 1
II. Needs for Maintenance of the Fluid Medium 5
III. Needs for Organic Structure 7
IV. Needs for Structural Organization 10
V. Needs for the Catalysis of Organic Transformations 13
A. The Mobile Cations 15
B. The Halogens 18
C. The Heavy Metals 19
Summary and Comment 25
References 25

Metabolism of the Major Mineral Elements (Ca, P, Mg, S and also Fe) in Relation to Requirements
D. P. CUTHBERTSON

I. Role of Major Minerals in Marine Animals 28
A. Introduction 28
B. Phyletic Review of Primitive Groups in Equilibrium with Sea Water 29
1. Decapod Crustaceans 29
2. Mollusca 29
3. Tunicates 30
4. Fishes 30

C. Major Minerals in Tissues and Skeletons 30
1. Eggs, Embryos and Larvae 31
2. Skeletons of Marine Animals 31
D. Factors Influencing Requirements of Aquatic Animals 32
E. Marine Birds and Mammals 33
1. Marine Birds 33
2. Marine Mammals 33
II. Requirements of Fish 34
III. Insects 36
IV. Metabolism of Calcium and Phosphorus in Relation to Requirement 37
A. Calcium and Phosphate Metabolism 38
B. Calcium and Phosphate Absorption 39
C. Calcium and Bone 42
V. Metabolism of Magnesium in Relation to Requirements 44
A. Magnesium Deficiency 46
1. Interrelationship between Body Temperature and Magnesium Concentration in the Blood 46
B. Excretion 47
VI. Metabolism of Sulphur other than as Sulphur-Containing Amino Acids 48
A. Sulphated Compounds 49
a) Sulphated Polysaccharides 49
b) Chondroitin Sulphates 49
c) Mucoitin Sulphates 49
B. Iron-Sulphur Proteins 49
C. Heparins 50
D. Thiocyanates and Isothiocyanates 50
E. Thiamin (Vitamin B_1) 50
F. Sulphide Sulphur 51
G. Sulphate Sulphur 51
VII. Metabolism of Iron in Relation to Requirements 51
A. Haemoglobin 51
B. Chlorocruorin 52
C. Haemoerythrin 52
D. Metabolism 53
1. Absorption 53
2. Transport 53
3. Storage 54
4. Excretion 54
E. Iron Requirements 54
VIII. Factors Influencing Requirements of Domestic Birds 58
A. Calcium and Phosphorus 58
1. Growing Poultry 59
2. Laying Hens 60
3. Turkeys 60
B. Magnesium 61
1. Growing Poultry 61

C. Iron 61
IX. Factors Influencing Requirements of the Major Elements by Mammals – Calcium, Phosphorus, Magnesium, Iron and Sulphur 62
A. Major Minerals in Milk of Different Species 63
B. Calcium 64
1. Accretion of Calcium during Growth 64
2. Deposition of Calcium in Growing Animals 65
3. Calcium Deposition in Pregnancy 65
a) Deposition of Calcium during Pregnancy in Sheep 65
4. Calcium Excretion in Lactation 66
X. Requirements of Ruminants for Calcium, Phosphorus and Magnesium 66
A. Pregnancy 66
B. Lactation 69
C. Antlers 70
D. Sulphur in Ruminant Nutrition 70
XI. Factors Affecting Requirements of Horses for Major Minerals 71
XII. Requirements of Laboratory Animals 72
XIII. Requirements of Pigs 73
XIV. Requirements of Subhuman Primates for Major Minerals 74
A. Calcium and Phosphorus 74
B. Magnesium 75
C. Iron 75
XV. Requirements of Man 75
Summary 78
References 79

Trace Elements in Animals
R. B. WILLIAMS

I. General Considerations 87
A. Nomenclature 87
B. Occurrence 89
II. Qualitative Requirements of Different Animals 93
A. Invertebrates 93
1. Protozoa 93
2. Insects 94
3. Worms 96
B. Vertebrates 97
1. Mammals 97
2. Birds 97
3. Fish 98
4. Amphibia and Reptiles 98
III. Functional Requirements 99
A. Differentiation and Pre-Natal Development 99
B. Growth and Post-Natal Development 100

C. Reproduction (and Lactation) 103
IV. Effects of Deficiency and Toxicity 104
V. Metabolism and Biological Role 117
Summary 123
References 123

Activity Spectrum of Ingested Toxicants

T. D. LUCKEY

I. General Consideration 144
A. Nutrition Orientation 144
B. The Continuum of Hormology, Nutrition and Toxicology 148
C. Ingested Toxicants 149
1. Essential Nutrients 150
2. Antimetabolites 152
3. Food Toxicants 154
4. Other Toxicants 158
II. Phylogenetic View 160
A. Protozoa 165
B. Diploblasts 168
C. Acoelomates 168
D. Pseudocoelomates 169
E. Schizocoela 170
F. Entercoela 172
III. Adaptation 173
Summary 174
References 175

Subject Index 179
Index vol. 1 208

Preface

Following the pattern set initially, the second volume of this series offers comprehensive and authoritative reviews of the present knowledge of nutrition from a comparative point of view. The present volume focuses on the chemical needs of animals, covering both major and trace elements as well as the effects of toxicants.

From the material reviewed therein it is evident that some general principles with regard to chemical needs of animals are beginning to emerge, such as the existence of an inverse relationship between the atomic numbers of the elements and their abundance in the universe. Over 99% of the structures of animals are derived from 13 of the first 20 elements in the periodic table, while 8 of the next 10 elements serve obligately in a biocatalytic capacity. Virtually all the elemental needs of animals were acquired in the relatively constant marine environment of the primordial stage of development. Bertrand's law for plants applies equally to animals, in that life is impossible with a deficiency of essential elements, but an excess of these is toxic. While essential elements may be toxic at high doses, many other ingested toxicants act as stimulants in minute quantities. The latter phenomenon, which is comparable to the growth stimulation obtained by feeding antibiotics or other growth promotants, forms the basis for considering the complete dose-response relationship as a continuum. This blending of nutrition, hormology, and toxicology presents a unified concept of the intricacies by which ingested chemicals may cause opposite and diverse reactions depending upon the dose, the conditions under which they are used, and the character of the individual exposed to them.

The Editor is indebted to the individual authors for their cooperation and to the S. Karger AG staff, particularly Mr. THOMAS KARGER, Mr.

R. STEINEBRUNNER and Miss A. ROGAL, for their guidance and support. A special note of thanks is due to Miss KAREN MARIE RECHCIGL for her assistance in the preparation of the index.

Washington, D.C.
October 1976

MILOSLAV RECHCIGL, Jr.

Abbreviations and Symbols

ADP	Adenosine 5′-diphosphate
AMP	Adenosine 5′-phosphate
ANTU	α-Naphthylthiourea
ARC	Agricultural Research Council (UK)
ATP	Adenosine 5′-triphosphate
ATPase	Adenosine triphosphatase
CoA	Coenzyme A
CT	Calcitonin
CTP	Cytidine 5′-triphosphate
DDT	1,1,1-Trichloro-2,2-bis-(*p*-chlorophenyl)-ethane
DM	Dry matter
DMBA	9,10-Dimethyl-1,2-benzanthracene
DNA	Deoxyribonucleic acid
2,4-D	2,4-Dichlorophenoxyacetic acid
ECF	Extracellular fluid
EDTA	Ethylenediaminetetraacetate
EPA	Environmental Protection Agency (US)
FDA	Food and Drug Administration (US)
GDP	Guanosine 5′-diphosphate
GTF	Glucose tolerance factor
Hb	Hemoglobin
HC	Hydrocarbon(s)
5-HT	5-Hydroxytryptamine
LD_{50}	Lethal dose for 50% of test subjects
LD_{100}	Lethal dose for 100% of test subjects
ME	Metabolizable energy
MRC	Medical Research Council (UK)
MYL	Anti-louse powder
NAD	Nicotinamide-adenosine dinucleotide
NADH	Nicotinamide-adenine dinucleotide (reduced)
NAS	National Academy of Sciences (US)
NPN	Non-protein nitrogen
NRC	National Research Council (US)
NTA	Nitrilotriacetic acid
PTH	Parathyroid hormone
RDA	Recommended daily allowence(s) (US)
RM	Receptor-metabolite complex
RMI	Receptor-metabolite-inhibitor complex
RNA	Ribonucleic acid
TCA	Tricarboxylic acid
TDE	Tetrachlorodiphenylethane
TFM	3-Trifluoromethyl-4-nitrophenol
UDP	Uridine diphosphate
UTP	Uridine triphosphate
ZEP	Zero equivalent point

Comp. Anim. Nutr., vol. 2, pp. 1–27 (Karger, Basel 1977)

The Chemical Element Needs of Animals

ERNEST BEERSTECHER

Department of Biochemistry, Dental Branch of the University of Texas at Houston, Houston, Tex.

I. Introduction

Any consideration of the elemental needs of animals must be accepted as a progress report on philosophy and experimental technology, rather than as any conceptual fulfillment in an area of biological science. Five elements have been added to the list of chemical essentials in as many years prior to the date of this writing (1974), and another 20 elements are presently under intensive investigation as likely candidates for that list. To some extent this is due to remarkable improvements in technology. Almost equally important, however, are bold and imaginative advances in conceptualization which, for instance, lead the vision of the order of magnitude of chemical needs down some four or five times, to the threshold where only one atom of a specific element would be present for one specific gene in the chromosomes of a cell [SCHWARTZ, 1974].

Some general principles with regard to chemical needs are now beginning to emerge from the tenuous thread of knowledge, which begins with the observations of SYDENHAM and WILLIS on iron in the 17th century, and the studies by CHATIN on iodine reported in 1850. These have provided meaningfullness and guidance to investigators in recent years, and now point with some clarity to the direction of things to come. They may thus be of some value in the present ephemeral report.

Firstly, there exists an inverse relationship between the atomic numbers of the elements and their abundance in the universe and generally in the biosphere. The life process depends upon availability of elements in the environment, and it may be observed that over 99% of the structure of living things is *derived* from twelve (eleven in animals) of the first 20 elements in the periodic table (table I). Thirteen of the first 20 are known to be *essential* for animals, boron for plants at least, and lithium, beryllium and aluminium are under close scrutiny in this regard. The other three of the first 20 are inert gases.

In the first series of transitional elements, vanadium through zinc (No. 23–30) represent a continuous sequence of eight elements known to be essential for higher animals and thus a locus of trace element function. Beyond this group, only selenium, molybdenum, tin and iodine (No. 53) are known to be needed, and only cesium (No. 55) appears at this time to be a likely candidate for discovery among subsequent elements. Thus, all of the known chemical element needs of animals fall within the first half of the periodic table.

Secondly, virtually all of the elemental needs of animals were acquired in the relatively constant marine environment of the primordial stage of development. It is speculated that the bulk elements were incorporated into the vital process first, later iron and copper as more elegant systems of metabolism evolved, and only long afterward the remaining trace elements to serve more sophisticated functions. The evolutionary needs for elements must originally have become fixed long before the Jurassic when the first terrestrial vertebrates left the litoral environment. The needs for iodine and possibly chromium were among the last to appear. The new terrestrial environment met the established chemical requirements of animals in an irregular fashion. There was a relative deficiency of chlorine, iodine, selenium and molybdenum, and an irregular distribution of cobalt and molybdenum. Excesses of other elements also existed. Thus, more efficient homeostatic and excretory systems evolved. Significant chemical imbalances at this stage of evolution probably were countered by evolutionary change.

Manganese is excreted in the bile and pancreatic juices [COTZIAS and PAPAVASILIOU, 1964], and the antiquity of the hepatopancreatic anlage suggests that this element had major biological functions in very early invertebrates. The evolution of the kidney to conserve the more sparse fresh water concentrations of elements came much later in order to meet the needs of the already fixed marinelike composition of the extracellular fluids. It is thus of interest that the cationic trace metals are excreted by the gut, whereas the alkali metals, alkaline earths, and anions are excreted by the kidney [SCHROEDER, 1965].

Thirdly, Bertrand's law for plants applies equally to animals, in that life is impossible with a deficiency of essential elements, but an excess of these is toxic. Thus, the highly evolved animal can only survive in an environment

Table I. Periodic distribution of element requirements of animals

Representative elements								Transitional elements									
1	2																
H	He																
3	4																
Li	Be																
		5	*6*	*7*	*8*	*9*	10										
		B	*C*	*N*	*O*	*F*	Ne										
11	*12*																
Na	*Mg*																
		13	*14*	*15*	*16*	*17*	18										
		Al	*Si*	*P*	*S*	*Cl*	Ar										
19	*20*							21	22	*23*	*24*	*25*	*26*	*27*	*28*	*29*	*30*
K	*Ca*							Sc	Ti	*V*	*Cr*	*Mn*	*Fe*	*Co*	*Ni*	*Cu*	*Zn*
		31	32	33	*34*	35	36										
		Ga	Ge	As	*Se*	Br	Kr										
37	38							39	49	41	*42*	43	44	45	46	47	48
Rb	Sr							Y	Zr	Nb	*Mo*	Tc	Ru	Rh	Pd	Ag	Cd
		49	*50*	51	52	*53*	54										
		In	*Sn*	Sb	Te	*I*	Xe										
55	56																
Cs	Ba																

which provides the essential elements within concentration limits to which its absorptive and excretory capacities are adapted. That point of more-or-less final evolution was at least partially established by the fact that the absorption of trace metals from the environment is inversely related to the concentration of environmental calcium. This important relationship may be observed in the numerous studies that show metal ions to be far more toxic to aquatic life in soft water than in hard water, and in the relationships of copper, manganese and zinc deficiencies in domestic animals to the calcium content of the diet.

The French naturalist LAMARCK first called attention to the *biosphere* as a distinct portion of the lithosphere in the 18th century, but it was not until the work of VERNADSKY [1945] that the *noosphere* received attention as that stage through which the biosphere is now passing, and in which it is being modified and reconstructed by human effort [BEERSTECHER, 1954]. During this period of some thousands of years, agricultural and technological practices have modified the habitable face of the earth in considerable measure. Soils have been depleted and pastures have been over-grazed, and the resulting impoverishment has been translated directly into deficiencies of essential elements in animals. Other elements, essential or not, have been taken from subsurface mineral deposits and introduced into the living environment. These appear in the surface soil, water, air, and ultimately in food itself. During recent decades, this process has accelerated tremendously, and SCHROEDER [1965] has pointed out that urban air samples have been found to contain measureable quantities of some six essential and seven nonessential or toxic metals. Many of these tend to accumulate in animals with age.

Were the consequence of this latter consideration simply one of the toxicity of certain trace elements, the subject would not be one of great nutritional importance. The broad and poorly explored area of metal interaction in biological systems, however, provides new problems in assessing the absolute chemical needs of animals. In some instances, an element that is nonessential in the usual sense may be present in such amounts as to stimulate the effect of or even to substitute for an essential element, thus leading to erroneous conclusions as to its own significance, or masking the existence of the actual functional element. On the other hand, antagonisms between metallic ions may greatly elevate the requirements of some substances in order to achieve their biological role, in some manner analogous to that in which calcium antagonizes trace element absorption. It has been recognized for some time that a given dietary level of copper, for instance, may yield either deficiency or toxicity symptoms, depending upon the accompanying levels of molybdenum and sulfate, or zinc and iron. The fact that trace element deficiencies and excesses produce such a variety of clinical and pathological disorders only emphasizes that a tremendous number of these interrelationships remain to be explored [UNDERWOOD, 1971]. The direction of future trace element research may in fact be determined by man-made trace element redistribution in the environment, as well as the changing habits of man which so thoroughly influence his nutritional status [MERTZ, 1974].

It is apparent from the foregoing considerations that the criterion for the essentiality of an element can no longer rest on the experimental demonstration that life is impossible in its absence. The stoichiometric relationship of an element in association with enzyme activity is generally accepted as a certain indication of its essentiality, but this approach involves distinct analytic limitations. At this point in time, the most practical definition of a chemical need is that an element is essential if its deficiency reproducibly produces an impairment in function and performance [MERTZ, 1969]. When any of an almost endless variety of deficiency symptoms can be produced and cured by diets containing varying concentrations of an element, there must be considered to be strong presumptive evidence for its essentiality.

A more detailed consideration of the elemental needs of animals may most profitably be undertaken in the light of the chemical functions performed by each. The discussion that follows is therefore organized as to needs for maintenance of the fluid medium, for organic structure, for structural organization, and for the catalysis of organic transformations. As a consequence, it will become apparent that the rigid economy of the periodic table has often adapted the use of a single chemical element to fulfill more than a single function, and that these multifaceted needs arose at quite different periods in the evolutionary process.

II. Needs for Maintenance of the Fluid Medium

Quantitatively, the major requirement for minerals in animals is caused by the need for maintenance of the internal environment in an osmotic and electrolyte balance established in primordial marine forms. Only in a few select groups of lower species has this tremendous dependence been partially removed by the adoption of osmotically active organic solutes that can be produced endogenously by metabolic processes, much as bicarbonate is supplied in this fashion generally. In the elasmobranchs, metabolically derived urea and trimethylamine levels of the plasma appear to be significant in osmoregulation. Free amino acids in insect hemolymph act for the most part as cations, and range in concentration from 300 to 2,400 mg%, as contrasted to levels of 50–100 mg% in higher animals. Blood sugar (trehalose) levels in insects commonly range from 1 to 6%, thus exerting considerable osmotic activity. In some species, glycerol levels may increase to as high as 25–30% of the entire insect, providing major advantages in cold-hardiness.

In this remarkable set of adaptations, metabolic acids (lactate, ketoglutarate, citrate, succinate, malate, fumarate) account for a considerable portion of the anion level – in *Gasterophilus* larvae as much as 50% of the total. Chloride is seldom the predominant anion in these lower animals. Primitive and zoophagous insects tend to have high sodium requirements and high hemolymph concentrations, while later phytophagous species that developed along with the angiosperms have higher potassium and magnesium levels and requirements. More intimate relationships between phylogeny and dietary requirements are apparent among the insects than in any group of higher animals studied to date.

With the foregoing exception, osmotic and electrolyte balance in animals is attained by fulfillment of the chemical needs for sodium, potassium, calcium, magnesium, chloride, and phosphate. Relatively smaller amounts of calcium, magnesium, and phosphate are required for structural purposes, and minute amounts of all for catalytic functions in various systems. Sodium is invariably the major extracellular cation, and along with calcium must be balanced against the intracellular levels of potassium and magnesium. In a similar fashion, extracellular chloride and bicarbonate are matched against cellular phosphates, sulfates, and protein. The relationship between the nutritional needs for these elements is quite narrowly fixed for any given species. The relatively large amounts of calcium and magnesium stored in the endoskeleton of the vertebrates may account for the lower levels of alkaline earths found in their fluids, and thus required. The fluids of invertebrates contain a much lower ratio of (Na + K) / (Ca + Mg) than do those of vertebrates, and this is a rather absolute consideration, independent of diet and other phylogenetic considerations. For vertebrates in general, the ratio is over 12 (21.7 in man), while in the invertebrates it is invariably below 7 [FLORKIN, 1949]. Variations such as these must be considered as evolutionary compensations to maintain rigid chemical interrelationships of a more universal nature in animal cells at the organellar and physicochemical level. At this irreducible and fundamental point, the major dietary requirements for these elements may be seen as reflections of needs for the stabilization of cell walls, the initiation of extracellular reactions and the support of energy-driven transport system [WILLIAMS and WACKER, 1967].

It may appear superfluous to state that the concentration of the fluid environment is maintained within narrow limits by the balance between water intake and loss, and that a variety of evolutionary devices has appeared to regulate this loss. The quantitative need for exogenous water is not absolute, however, since under conditions of acute water shortage, endogenous metabolic water may meet a portion of the requirement.

During water deprivation, the metabolism of fat may produce over 100 g of water per 100 g of fat oxidized, twice as much as is produced from carbohydrate or protein, and without appreciable ketosis. The need for the maintenance of the bicarbonate ion in the internal milieu is continuously met by the endogenous metabolism of organic materials. In this instance, of course, the anion is continuously lost through exhalation, and virtually all reduced carbon compounds entering the animal organism must be considered as ultimate sources of this critical inorganic material. Along with the other ions that regulate the osmotic pressure, bicarbonate maintains the pH of the fluid medium. For land animals, the hydrogen ion concentration so achieved generally falls quite close to a pH of about 7.4; a wider range of about 7.2–7.8 is seen in marine animals. Insects uniquely exhibit an internal medium with a pH of about 7.0 or less, a fact of considerable importance in delineating the ionic behavior of the high free amino acid levels in their fluids.

III. Needs for Organic Structure

Six of the lighter elements of the periodic table are required in the animal economy in relatively large amounts as components of their predominantly organic structure: carbon, hydrogen, oxygen, nitrogen, sulfur, and phosphorus. With the exception of oxygen, none of these may be utilized in elemental form, and all except phosphorus are unavailable to the animal economy except in the combined forms represented by a very few organic alcohols, amines, acids, sulfides, and their derivatives. As a concession to their evolutionary departure from the autotrophic state, moreover, all animal cells absolutely require for their substance a small number of specific amino acids. A much greater variety of amino acid type compounds, carbohydrates and lipids serve the major portion of organic structural raw materials. Scarcely a decade ago, the problems of animal nutrition seemed largely to center around the conversion of readily available plant materials to meet animal needs for these elements. More recently, however, perspectives have given rise to the recognition that the photosynthetic mill is not inexhaustible, and that the rapid pace of technical and social change may even provide a threat to this primordial *sine qua non* of chemically reduced carbon and nitrogen compounds.

Molecular oxygen and *water,* which had established themselves in measurable quantities upon the earth some two billion years ago, were to provide the special chemical pattern for eventual life as the earth's heat approached present levels at about that same time. When the earth was first formed, its chemical composition was much the same as that of the sun. The sun's protosphere contains some 81.76 atomic percent of hydrogen, however, as against 2.7 atomic percent of this lightest element in the earth's crust. The difference is accountable to diffusion away of the hydrogen, along with a concurrent loss in the earth's mass and the concentrations of its heavy gases (by distillation), xenon and krypton. Free oxygen was similarly lost in this process, so that the present high concentration (48.7 atomic percent as against 0.03 for the sun) is attributable to water vapor that was released from the interior of the earth during its long cooling period. Water vapor below the tropopause underwent photolysis, and consequent oxygen enrichment was further achieved by continued hydrogen diffusion away from the earth. The fact that molecular oxygen absorbs solar radiation in the very range that is photolytic resulted in an eventual decrease in the dissociation of water vapor and hydrogen loss.

Thus arose the virtually unique oxidizing atmosphere upon the earth which has maintained itself for several billions of years in a relatively constant state and permitted the evolution of higher life forms. Acid-forming elements, including carbon, reacted with silicate rocks in such proportion as to provide minute atmospheric concentrations. The maintenance of this status most evidently rests on the constancy of the light-transmitting properties of the atmosphere itself. The total amounts of oxygen, carbon dioxide and water that are involved in the life process upon the earth are enormous, and small external changes in the nature of radiation reaching the surface of this isolated planetary ecosphere could well provoke severe long-range distortions in balance. Earthwide monitoring of radioactive fall-out has dramatized the reality of atmospheric pollution being far broader than a local problem, and man-made pollution continues to increase at an astonishing rate. Of further and growing concern is the amount of change in the composition and stratigraphy of the upper layers of the atmosphere through the intervention of aircraft. The ability of man to create fatal, long-range changes in the planetary economy is thus no longer unforseeable.

Carbon in the form of carbon dioxide is a rare gas in the atmosphere, accounting for only about 0.03%. In this state it is maintained by demands for photosynthesis and by the reservoirs of carbonates dissolved in surface waters and in carbonate rocks. The annual photosynthetic demand has been estimated at about 150 billion tons; of this, some portion is removed from the continuing plant-animal metabolic cycle and geologically sequestered in the form of coal and petroleum. It is seldom mentioned that a major portion of the photosynthetically produced organic carbon (e.g. cellulose) is solely available to meet plant requirements (excepting in a few symbiotically adapted animal species). Another huge portion of the reduced carbon is similarly denied the higher animals by its invertebrate incorporation into nutritionally unavailable chitin. Added deleterious factors may thus make nutritionally available carbon sources difficultly obtainable on the earth now as it has been throughout man's development.

As a result of these considerations, the use of unconventional and synthetic energy sources for animal nutrition have come under increased attention. Microbial products – grown from such materials as petroleum hydrocarbons and even methane – have been of some limited interest in this regard. Unconventional synthetic nutrients are now appearing as a subject of increasing research. 1:3-Butanediol, for instance, when fed at the 5% level (isocalorically), has been found to provide a metabolizable source of carbon [KIES *et al.,* 1973], being utilized via the pathways of alcohol dehydrogenase and fatty acid oxidation. These observations suggest enormous possibilities for future nutritional carbon sources.

Nitrogen is required by animals in reduced organic combinations, a very limited number of which are biologically acceptable. A few specific organic combinations (essential amino acids, vitamins) are absolutely required by all animals. It is of interest that the nutritional requirement for purines and pyrimidines seems *inversely* related to phylogenetic development. Nitrogen deprivation of animals is related to the low protein content of much plant material, and the highly irregular plant providence of certain of the essential amino acids. Nitrogen is largely eliminated from animals in reduced organic form, also, and its various transformations back to acceptable animal material are almost entirely performed by bacteria. Bacterial reactions include nitrogen fixation, oxidation, reduction and conversion to organic forms, the latter process being augmented by higher plants. So dynamic is this process that, without nitrogen-producing bacteria, nitrogen-fixing bacteria would at their present rate remove all nitrogen from the atmosphere in a few million years. Except for chitin-splitting bacteria, atmospheric nitrogen would all be converted to chitin in about 40 million years.

As an alternative to biologically fixed nitrogen, urea has been used for some time as a source of nonspecific nitrogen in ruminants. More recently, it has been found to be utilized to some extent by nonruminants and man [KIES *et al.,* 1965]. Subsequently, it has been found that even ammonium salts are capable of improving nitrogen balance in man under certain experimental circumstances [FÜRST *et al.,* 1969], and it may be anticipated that a growing list of synthetic nitrogen sources for higher animals will gradually evolve.

Sulfur in the form of sulfates is sufficiently ubiquitous to meet the nutritional needs for this element by plants, but its obligate requirement in the form of methionine by animals may cause severe limitations to animal growth. The extensive use of sulfur derivatives in higher animals (as contrasted to plants) for keratin and for chondroitin mucopolysaccharide structure puts methionine in a particularly critical key role in animal nutrition, and one that may only find ultimate solution in industrial methionine production. Sulfur is also required in animals in organic combinations in biotin, coenzyme A, lipoic acid, and thiamine, although bacterial symbionts frequently are effective in producing these sulfur combinations.

Phosphorus in the form of its phosphates is also sufficiently ubiquitous that its deficiencies are rare and unknown to affect its principal role in organic structures in the energy transfer systems, and in nucleotide structure. Unlike the other major elements discussed in this section, it is unknown to function in reduced form in animal systems.

IV. Needs for Structural Organization

With the evolution of multicellular forms of life and the growth of organisms having dimensions many times those of their component cells, additional chemical requirements arose to provide form and rigidity to higher species. Nature appears to have experimented with the problem at the single-cell level, for skeletal protective coverings of the broadest variety may be seen in both plant and animal species. Connective tissue supportive structures *common to both* plants and animals evolved to meet the needs of more complex species, in keeping with the principle that when nature 'finds' a good way to do something – to build a structure to carry out a certain function, or to create a process to supply a certain need – then that same basic plan will be found to be used time and again, in both the plant and animal kingdoms. [PERSON, 1968]. Supportive structures thus arose in both kingdoms which consisted of organized microfibrils embedded in a ground substance containing large amounts of acidic polysaccharides. The fibrillar component *in higher animals* most generally is protein (*e.g.* collagen and elastin); it may be chitin in fungi and arthropods, however, and in the tunicates and in plants it is largely cellulose. The molecules of polysaccharide that characterize the ground substance all have similarities in composition and structure, as well as common pathways in their biosynthesis [MATHEWS, 1968]. Of tremendous significance in this latter regard is the universal utilization of nucleotide sugar phosphates as sugar donors. It is around this common organic matrix that the mineralized supportive structures of higher animals have subsequently evolved.

Silicon in the form of silicates has been recognized for a very long time as beneficial for certain forms of life, if not essential. Thus, a considerable number of flagellate and amoeboid protozoa have siliceous tests, and these have been of great significance in micropalaeontology. Neither the nutritional requirement for silicates nor the metabolic process by which they are deposited are known, although it has been presumed that an organic matrix for their crystallization is essential. No more is known concerning the siliceous skeletal deposits of some sponges. It is also well recognized that silicates play some role in providing rigidity to the cellulose fibers of higher plants. The stems of wheat, for instance, are unable to stand up in a mild breeze, and are more susceptible to parasitic destruction when the plants are grown in a silicate-deficient medium. In studies attempting to elucidate the biological role of silicon, it has been found that this element tends to be concentrated in connective tissues, and stoichiometrically in the glycosaminoglycans of animals and polyuronides (pectins and alginates) of plants. Silicates now appear to function as structural cross-linking agents in connective tissue polysaccharides. This is apparently also true in some proteins, such as collagen [SCHWARTZ, 1974]. The silicon atom is small and has the same stereochemistry as carbon. It readily forms thermodynamically very stable Si-O-C bridges. In its natural abundance in the biosphere, it is second only to oxygen. The crucial role of silicon in animals and plants is thus in keeping with its chemical nature and distribution.

The essentiality of silicon for growth in rats was observed by SCHWARTZ and co-workers as early as 1965 in an all-plastic isolator system, but the relatively large amounts required led them initially to suspect trace element contamination of the 50 mg of sodium metasilicate per 100 ml in the diet. Silicon was subsequently found to be essential for both growth and the prevention of bone deformities, and these observations were simultaneously made by CARLISLE [1970a] working with chicks. Electron probe microanalysis has also shown [CARLISLE, 1970b] that as much as 0.5% of silicon may be found in very narrow regions of active calcification in the area of bone growth, suggesting a possibly separate role in mineralization [CAR-

LISLE, 1974]. While the subsequent observation of the biological role of silicon as a cross-linking agent in macromolecules provides a reasonable explanation of the deficiency symptoms, it does not preclude the existence of still other possibly catalytic functions of this element; indeed, silicon has long been implicated in some fashion in iron transport in green plants. Silicon apparently antagonizes the metabolism of fluoride in some fashion [MILNE and SCHWARTZ, 1974], and fluoride is involved in iron-pigment deposition in teeth.

Fluorine has for half a century been recognized as beneficial in the prevention of dental caries and in the maintenance of normal skeletal development, but the essentiality of the element in the conventional sense has not been established until recently [SCHWARTZ, 1974]. The role of fluorides in mineralized tissues has thus been considered as a structural one, involved with the benefits to be derived from the physical properties of the fluoroapatite crystalline lattice. The clinically optimal level of about 1.5 ppm in drinking water is significantly the same as that occurring in sea water, an observation that beclouds the fact that fluorine is more abundant on the earth's surface than is chlorine! The relatively high toxicity of fluoride, moreover, has somehow tended to discourage investigation of the possibility that this element might be more generally essential at similar levels in a catalytic fashion. Rats have been shown to require as much as 2.5 ppm of fluoride in their diet for optimal growth and normal incisor pigmentation. The normal orange-brown pigment found in the upper layers of dental enamel is missing in fluoride-deficient animals, and significant improvement in pigmentation occurs with subsequent fluoride supplementation of the diet. Higher than physiological levels of fluoride again cause enamel depigmentation, presumably due to the formation of the highly stable $(FeF_6)^{-3}$ complex (mottled enamel).

The growth promoting effects of added fluoride for rats on a properly balanced deficiency diet strongly suggest a catalytic role for this element. Fluoride has been reported to stimulate growth in tissue culture, to activate citrulline synthesis by liver and to activate adenyl cyclase. It has been pointed out that mammalian regulation of the blood and tissue levels of this element in itself suggests the essentiality of fluoride. Excesses are rapidly excreted or deposited in bone, which further acts as a reservoir to maintain a constant 0.1–0.2 ppm fluid level. The placenta accumulates fluoride and regulates its transfer to the fetus, and constant levels approximating those of the body fluids are found in milk. Similar degrees of homeostasis are seldom associated with the nonessential chemical trace elements.

Calcium-containing deposits are often associated with the exterior of many bacteria, algae and fungi. These are often, but not always, precipitates. Highly organized calcified structures are common with *Protozoa,* however, and as in the higher animals, these are invariably deposited upon an organic matrix. The composition and degree of mineralization of these supporting structures vary greatly from species to species. Strontium sulfate is said to compose the tests of *Acantharia.* In most *Protozoa,* however, calcium carbonate is the major ingredient of the test, although phosphates do occur in a few. A similar consideration holds with the sponges, while phosphates tend to appear in somewhat greater quantity in the corals. Approximately 90% of the mollusc shell is composed of calcium carbonate, while in the crustacean exoskeleton, the percentage of calcium phosphate in the mineral material approaches the levels found in vertebrates. Prior to molting in the crustaceans, calcium is transferred from the exoskeleton and stored in the gastroliths or hepatopancreas. This is redeposited into the exoskeleton when molting is complete, but since the process is not highly efficient, there is a very high chemical demand for dietary calcium at this latter time. Little information has become available as to the quantitative calcium and phosphate needs of the invertebrates, although it is observed that there is marked inhibition of skeletal formation in the sand dollar *(Echinarachnium parma)* when the levels in sea water are reduced to less than one fifth of normal (less than 80 mg/liter).

Approximately 99% of the calcium, 80% of the phosphorus, and 70% of the magnesium in vertebrates is found in the form of the hydroxyapatite crystals of bone and teeth. Aside from these needs, considerable amounts of sulfur (sulfate) are needed to synthesize both the mineralized and unmineralized chondroitin sulfates that characterize some parts of the ground substance. A large body of literature has developed concerning the mineral requirements for skeletal formation in vertebrates, as well as the numerous factors which influence this chemical need, and much of this is discussed in later pages of this monograph.

V. Needs for the Catalysis of Organic Transformations

Of the 25 chemical elements presently recognized as needed by animals, 14 are required in sufficiently small amounts that they are classed as trace elements. It is presumed that these function catalytically, although it is now recognized that silicon and fluorine have critical structural functions, and others may eventually be found to possess similar kinds of activity, perhaps in microcosms as yet undefined.

Of the so-called macroelements, Na, K, Ca, Mg, S (as sulfate), P (as phosphate), and Cl may also have catalytic functions in trace amounts, but these have not always been readily distinguishable. From a rigid point of view, hydrogen (ion) is the most generally active catalytic element in the living organism, and is present in only trace amounts in virtually all vital systems. It might be added that it is compatible with life in only a narrow range of concentrations. The ubiquitous and generous distribution of the macroelements would tend to make deficiencies of these materials at the catalytic level very difficult to observe in animal systems. In some instances, therefore, chemical needs have been demonstrable only in invertebrates or in plants, where needs for these to meet other functions as in higher animals may not exist.

Most of the elemental needs for catalysis were defined in the primordial biogenetic soup of the eobiont, and therefore by considerations of abundance and availability. Iron, which tops the heavy metals in its abundance in the earth's crust, is now known to occur in over 100 protein and enzyme combinations. The abundance of metals having atomic numbers above 50 is so low as to make them impractical for use in the life process. The high insolubility of the hydroxides of aluminum and titanium make these elements virtually unavailable, although they are abundant. The virtual absence of chromium and nickel from living material (excepting RNA) has been attributed to the remarkable stability of these cations in the octahedral sites of soil silicates.

Trace element *ions* are virtually nonexistent in the free state in living material, but exist as complexes with small organic molecules or macromolecules, depending upon their chemical properties and their biological functions. Since the heavy metal ions must of needs have more or less fixed chemical neighbors, it is essential that they have specific and properly distributed binding sites, and therefore quite selective organic chemical partners or *ligands*. The heavier metals tend to bind preferentially with the nitrogen or sulfur ligands of amino or sulfhydryl groups, for instance, while magnesium and calcium are more frequently bound to the oxygen ligands of phosphate or carbonate groups [WILLIAMS, 1967]. Depending upon the firmness of the association of the metals with their site of activity, three separate groups tend to emerge: (1) *The mobile cations.* Sodium and potassium are the most mobile cations and function in nerve conduction, while calcium and magnesium (the next most mobile) act in triggering muscle action. (b) *Metalloenzymes.* These substances dissociate at concentrations of 10^{-10} M. Well over 100 of these have been characterized, and the fundamental role of the metal ascertained. (c) *Metal enzyme complexes.* These dissociate more readily at concentrations of 10^{-4} to 10^{-5}M, allowing for a 100,000-fold distinction from the previous group. This loose association of the metal makes the study of elemental functional roles quite difficult.

The discussion which follows of the individual elemental needs for the catalysis of organic transformations proceeds by a consideration of the mobile cations and the halogens, and then the heavy metals in atomic order. Reference has already been made to the catalytic functions of fluorine and silicon (pp. 11, 12).

A. The Mobile Cations

Sodium is the classic example of an element that is so ubiquitous as to make studies of its function at the subcellular and catalytic levels exceedingly difficult. Since cells evolved in an essentially saline environment, it appears that most of the course of evolution has involved mechanisms for excluding the element from cellular interiors. Nevertheless, it would seem unlikely that some employment of this element at the subcellular level had not developed. As an approach to the examination of this problem, it seems necessary to consider the role of this element in plant life, in which the macro-roles of sodium are not so pronounced. It remains very difficult to demonstrate that higher green plants require sodium at any measureable level, but it is still not possible to rule out trace contamination of media with the element. By contrast, sodium is a distinct requirement of the algae studied. *Anabaena cylindrica* achieves optimum growth only in a medium containing about 5 ppm of the element. Neither potassium, lithium, rubidium, nor cesium are able to substitute for this chemical need, nor does sodium substitute for any of the potassium needed by this species. Deficient cultures do not grow as well and are impaired in phycocyanin synthesis, but chlorophyll synthesis is apparently unaffected. Sodium is said to be required by other algae in order to sustain high photosynthetic rates, but not for respiration. The sodium requirement of *Chlamydomonas* is elevated by potassium and depressed by magnesium. The necessity for a trace of sodium has been established in a number of bacteria and it is able to replace part of the potassium requirement of *Aspergillus niger*. Sodium is a known requirement of all insects studied, and is of course required in human tissue culture media. It must therefore be concluded that there is strong but presumptive evidence for its function in animals in trace amounts in catalytic transformations that remain to be elucidated with more recent sodium exclusion techniques.

Potassium is assiduously garnered by all cells and appears to be universally required in their nutriment. Both photosynthesis and translocation in plants depends upon its presence, and in potassium deficiency the leaves of higher plants turn to a dark blue-green to bronze color and manifest marginal chlorosis and necrosis. Potassium is known to be required by algae, bacteria, yeasts, and molds for growth and for spore formation, and in some bacterial species, the requirement can be partially replaced by rubidium and cesium. The chemical need in *Aspergillus niger* is sufficiently specific and the response so uniform as to permit the use of this species as an assay organism for potassium. The protozoa uniformly require potassium, *Tetrahymena geleii* achieving optimum growth only at levels of about 0.3 μmol/ml of medium. Although not widely studied among invertebrates, potassium is needed in trace amounts in most insect diets investigated, including various *Anopheles* species, *Theobalda incidens,* and *Pseudosarcophaga affinis.* Its requirement in trace amounts in vertebrate tissue cultures has long been well established. Balanced against calcium levels, it is intimately related to the maintenance of protoplasmic irritability levels.

Potassium has been shown to be a requirement in the glycolytic system for the enzyme pyruvic kinase, which catalyzes the conversion of phosphoenolpyruvate and ADP to enolpyruvate and ATP. Since the 2 mol of ATP formed per mol of glucose in this step are the sole net profit from the glycolytic process, and since the glycolytic system appears to be a universal one in living cells, it must be presumed that potassium on this basis alone is a universal catalytic requirement. There is evidence that potassium is also required in the fructokinase enzyme system, and that its catalytic functions may be more numerous than are now known.

Calcium has the fortunate chemical versatility of being relatively mobile or becoming fixed, depending upon the circumstances of its environment, in the form of carbonates, phosphates, sulfates, gluconates, carboxylates, and a variety of chelates. At a molecular level, this permits its regionalization and progressive transfer and transport through a wide spectrum of dissociable complexes, and adapts it well to catalytic purposes in biology. As a consequence, it is known to function as a co-catalyst in well over 100 enzyme systems, although these are, for the most part, systems that are peculiar to higher animals.

It appears that calcium serves catalytic functions in all levels of the plant kingdom, or at least is required in trace amounts. In higher green plants, deficiency levels produce poorly developed root systems, fruiting is impaired, and the leaves become curled, rolled and chlorotic. Algae, such as *Anabaena cylindrica,* have been shown to have a specific requirement for 20 ppm in the medium, none of which is replaceable by strontium or other elements. Among the bacteria and fungi, calcium deficiency causes impairment of growth and sporulation in *Aspergillus* and *Penicillium,* nitrogen fixation by *Azotobacter* and *Rhizobium,* nitrogen fixation by *Nitrosomonas,* hydrocarbon oxidation by *Pseudomonas,* and a long list of similar signs of deficiencies in others. It seems that calcium is required by the protozoa in trace amounts, but that the magnitude of the requirement is very low. The chemical need for calcium has been demonstrated convincingly in numerous insects, and for *Blatella germanica* around 50 ppm in the diet are required for optimum growth. Calcium is an essential requirement of all mammalian tissue culture media.

Some, but not all, of the activities of calcium are partially replaceable by other ions. For the most part, calcium functions in hydrolytic type reactions involving, for instance, phosphatases, lipases, maltase, and collagenase. It stimulates the ATPases of muscle protein, is critical in nerve impulse transmission, is essential to the clotting of blood and of milk, and indeed to milk production. Details as to its distribution at the subcellular level are now beginning to emerge to broaden the vision of its overall function in fundamental metabolism.

Magnesium is a cofactor for oxidative phosphorylation, and thus it can be said without equivocation that it is a universal nutritional requirement at the catalytic level. More generally, it is also necessary for the action of a large number of hydrolytic enzymes in all known forms of life, so that a long list of enzymes exists for which it serves as a cofactor. At a somewhat higher level quantitatively it functions as its nucleic acid salt. Magnesium is a structural component of the chlorophyll molecule, and thus is the lightest of the required elements to occur as a distinct metallo-organic metabolic cofactor. In its function as a catalytic element in hydrolytic reactions, it is frequently, at least partially, replaceable by manganese ions.

The magnesium requirement for green plants is relatively high because of the need for chlorophyll synthesis, and large amounts are also found in the plant connective tissue bound, along with calcium, to the ground substance pectins (p. 11). In plants, proper magnesium: calcium ratios in the nutritional source are critical, and the pectins may perform a limited type of homeostasis. Magnesium has added specialized functions in the bacteria, being essential for nitrogen fixation and in some instances for pigment formation. *Aspergillus niger* has been employed as a magnesium assay organism. A generally demonstrable requirement exists among the protozoa, *Tetrahymena geleii* being representative and requiring about 2 $\mu g/ml$ of medium for optimum growth. Small amounts have been shown to be crucial for the growth of insects, and in mammalian tissue culture. The requirements among higher animals are small for catalytic purposes, by contrast with needs for other functions, and the abundance of magnesium has not necessitated extended studies of exact needs for dietary purposes.

B. The Halogens

Fluorine has previously been mentioned (p. 12) in relation to its participation in hydroxyapatite structure, but the basis of its need for growth in the rat is unknown, as well as its function in iron transport. It is reported to be essential for hepatic citrulline synthesis and for the activation of adenyl cyclase. Fluoro fatty acids have been reported in nature and these are highly toxic to both cattle and humans.

Chlorine, as the chloride, has been shown to be essential for the activation of many amylases, but its activity is largely replaceable by bromide. Its requisite nature at the catalytic level has not been well established beyond this. It has been reported as necessary for several bacteria, including the halophilic species, and for *Blatella germanica* the need is somewhat less than 5 $\mu M/g$ of diet. Organic chlorocompounds are produced by a number of bacteria (chloramphenicol, chlortetracycline).

Bromine is present as dibromotyrosine in the protein gorgonin of some corals and also as dibromindigo in the pigment of certain gastropod molluscs. In higher animals and plants, the essentiality of bromine has not been established, partly because of its ubiquitous distribution, and partly because of its interchangeability to a variable extent with the chloride ion. The need for bromine, however, cannot as yet be discounted.

Iodine is required by higher animals for the synthesis of the thyroid hormones, thyroxine and triiodothyronine, and this is at present its only known function. It is present in some corals as diiodotyrosine in the protein iodonin or gorgonin. Scattered reports that it is needed by plants remain unconfirmed. Iodoamino acids are said to stimulate the growth of a number of invertebrates, but their essentiality has not been demonstrated. A growing number of naturally occurring organic goitrogens which interfere with thyroid hormone synthesis may produce conditioned deficiencies of iodine in both humans and livestock. Cobalt (and possibly manganese) deficiencies have also come under attention in this regard in both humans and farm animals [UNDERWOOD, 1971].

Mention of another nonmetal, *sulfur,* should be made at this point as it is involved in catalytic quantities in living cells in the form of various organic combinations which function as enzyme cofactors, such as biotin, coenzyme A, lipoic acid and thiamine.

C. The Heavy Metals

Vanadium (atomic No. 23) is the lightest of the presently recognized essential heavy metals, and its presence as a porphyrin complex in some ancient oil shales suggests that its ubiquity in the biosphere brought about its early utilization in life processes. The ability of vanadium to form coordination complexes and to exist in a number of valence states makes this element of particular adaptability to biological catalysis. It is generally present in all plants and animals and has a relatively low toxicity. Vanadium is apparently needed by higher plants, and absolutely required by lower ones, such as *Azotobacter* (for nitrogen fixation), and *Aspergillus niger.* It is found in very high concentrations in a variety of invertebrates, often in conjunction with high concentrations of sulfuric acid. A growing body of literature indicates that vanadium acts as an oxidation-reduction catalyst in lipid metabolism, but that this is not its sole function. There is evidence that it exerts a beneficial effect in preventing dental caries, but the biological significance of this observation is unclear. It is also involved in some manner in iron metabolism. Vanadium has now been demonstrated to be an absolute requirement for higher animals [HOPKINS, 1974; HOPKINS and MOHR, 1974], and in the chick and rat, deficiency produces reduced body and feather growth as well as impaired reproduction and survival of the young.

Chromium (atomic No. 24) has now been shown to be an essential element for a number of animal species, and human deficiencies are known to exist in certain areas in association with protein-calorie malnutrition. In addition, marginal human deficiencies are probable during pregnancy and old age. The major presently recognized function of chromium is as a type of cofactor for insulin, known as the glucose tolerance factor (GTF), which is released into the circulation whenever insulin is secreted or when a glucose load appears in the circulation. The element is active only in the trivalent state, which is converted, probably in the liver, into a particularly active form. The limited ability of animal tissues to synthesize GTF from chromium makes the necessity for analysis of foods for specific kinds of chromium an absolute necessity. Problems of analysis for this element have thus become of major importance, and appear to be completely insolvable until the chemical nature of GTF is further elucidated. The original observations on chromium occurred in relationship to the discovery that rats developed an impairment of intravenous glucose tolerance when fed on *torula* yeast in their diet, but not *brewers'* yeast [MERTZ, 1974]. This difference in yeasts suggests in itself some function of the chromium-containing factor in living cells other than that in higher animals.

Manganese (atomic No. 25) is apparently essential for all forms of life, including higher plants and bacteria. A strain of *Lactobacillus arabinosis* has, in fact, been employed as a manganese assay organism. The bacterial requirements for growth are appreciably below those for optimum function in synthesizing antibiotics, antigens, phages, enzymes, and spores, and a similar situation attains in higher animals because of the relationship of manganese to connective tissue formation, in addition to its basic role in intermediary metabolism. Its requirement has been demonstrated in various insects. Manganese deficiency occurs naturally in pigs and poultry maintained on certain diets, and has been demonstrated experimentally in a large variety of other domestic animals. It has not been observed in humans. Manganese occurs in at least six known metalloproteins, of which the most crucial appears to be pyruvate carboxylase. Interestingly, in manganese-deficient chicks, even the bound manganese of this enzyme is replaceable by magnesium. The protein avimanganin of avian liver (of unknown function) undergoes a loss of manganese content as well as a reduction in its own concentration under conditions of manganese deficiency, in contrast to the only other animal metalloprotein, the pyruvate carboxylase, which simply substitutes magnesium.

Manganese also serves in a very long list of metal-enzyme complexes, in which its action is not so specific (decarboxylases, hydrolases, kinases, and transferases). Many of the symptoms of manganese deficiency in higher animals are attributable to a breakdown in connective tissue formation, and more specifically to mucopolysaccharide synthesis (reflected in cartilage, bone and egg shell production, growth, reproduction, and blood clotting)

[LEACH, 1974]. Most glycosyltransferases are activated by manganese, and it is usually the most effective cation to activate these enzymes. The element is thus needed for the synthesis of hyaluronic acid, chondroitins, heparin, glycoproteins, such as prothrombin, and disaccharides, such as lactose.

Iron (atomic No. 26) is apparently needed by all living systems, and it has been proposed that, along with copper, it provided protection in the primordial development of the life process, against the highly toxic oxygen byproducts, superoxide ion and peroxide [FRIEDEN, 1974]. The situation has been dramatized by a heterotrophic, anaerobic strain of micrococci, which is only able to survive in the presence of oxygen when traces of heme are present. Iron is primarily involved in terminal oxidases and in respiratory proteins, both in heme and nonheme combination. Common heme systems include hemoglobins, myoglobins, hydroperoxidases, cytochromes, cytochrome oxidases, and tryptophan oxygenase. Nonheme iron proteins include the hemeerythrin oxygen transport system of invertebrates, the aconitase and succinic dehydrogenase of the Krebs' cycle, xanthine oxidase, and the nitrogenase involved in nitrogen fixation. While it has still not been conclusively demonstrated that all anaerobic bacteria require iron, requirements have been shown for many plants, and all invertebrates and higher animals investigated. It has been pointed out that at virtually every known step of iron metabolism, including storage, transport, biosynthesis, and degradation, the ferrous to ferric cycles play a dominant role in the multimetabolic pathways of the most important metal constituent in all living systems [FRIEDEN, 1973].

Cobalt (atomic No. 27) is essential to lower plants and presumably all animals as a portion of the vitamin B_{12}-containing enzymes, but its requirement by higher plants is sufficiently low that it has not been demonstrable at the present time. Some bacteria, molds, and algae are able to synthesize the vitamin B_{12} from elemental cobalt ions, while others require the preformed vitamin. Although animals normally subsist on supplies of vitamin B_{12} synthesized symbiotically by intestinal bacteria, nutritional deficiencies occur in domestic species in many areas of the world where cobalt is inadequate in the soil. This is particularly true in ruminants, in which the main source of energy is not glucose but acetic, propionic, butyric, and other fatty acids. In fatty acid utilization, methylmalonyl-CoA isomerase, which requires vitamin B_{12}, is essential for the isomerization of methylmalonate to succinate, and the demands for vitamin B_{12} are thus higher in ruminants than in monogastric animals. Vitamin B_{12} coenzymes also function in other reactions where a transfer of methyl groups occurs, as for instance in methionine synthesis. Conditioned cobalt (vitamin B_{12}) deficiencies are seen in humans in pernicious anemia, but a lack of cobalt is not apparent under other than these circumstances. Cobalt ions function in the activation of many other enzyme systems, but their effect is nonspecific. Cobalt is believed to be needed for nitrogen fixation by the root nodules of legumes.

Nickel (atomic No. 28) has recently been shown to be needed in the diets of rats and chicks at levels of several ppm in the diet to prevent changes in shank skin color and in liver consistency in the chicks, and reduced α-glycerophosphate oxidation by liver homogenates from both species [NIELSEN and OLLERICH, 1974]. Nickel has long been suspect as a trace element, meeting many essential chemical and metabolic criteria for such activity. *In vitro* studies show it to activate nonspecifically a number of enzymes. It stabilizes both RNA and DNA against thermal denaturation and enhances the adhesiveness of polymorphonuclear leukocytes. It may serve a role in the preservation of the compact structure of ribosomes, but these latter phenomena have not as yet been demonstrated *in vivo*.

Copper (atomic No. 29) is universally required by all known living species, including all types of plant life, invertebrates, and higher animals and, along with iron, was probably incorporated into the life process at a very early stage. Because of its reactivity, it is involved in a truly remarkable variety of enzymatic processes, and the symptoms which result from its deficiency in higher animals are therefore of the widest variety. Since early in the last century, copper was recognized as a component of the blood proteins of various invertebrates, and the red feather pigment of the South American turaco was found to contain as much as 7% copper. Copper deficiencies occur as grazing disorders of livestock, and as both frank and conditioned deficiencies in humans [UNDERWOOD, 1971]. Among the better-known copper-dependent systems are the hemocyanin oxygen carriers of invertebrates, the azurin and plastocyanin electron carriers in bacteria and plants, respectively, the superoxide dismutases, cytochrome oxidase in most cells, ceruloplasmin for iron mobilization, lysine oxidase involved in the cross-linking of collagen and elastin, tyrosinase involved in pigmentation, and galactose oxidase involved in sugar metabolism [PEISACH *et al.*, 1966]. As a result, deficiencies in higher animals are manifested by impaired growth and hematopoiesis, and defects in connective tissue formation, cardiac function, demyelination, pigmentation, keratinization, and bone formation. Copper is remarkably toxic to protistan plants and to invertebrates, and chronic copper poisoning in grazing animals is well recognized, being particularly common in sheep. The copper nutrition of grazing animals has shown itself to present particularly complex problems associated with the interrelationships of copper, molybdenum and sulfate in the diet, and the relationship of copper intake to zinc, iron and calcium in the diet.

Zinc (atomic No. 30) is a universal nutritional requirement for all forms of life, and its requirement by *Aspergillus niger* was demonstrated as early as 1869. Subsequently, it was found to be highly concentrated in a number of invertebrate tissues, part of the hemosycotypin respiratory pigment of certain snails, and essential for plant life. Among the more critical zinc-containing metalloenzymes are carbonic anhydrase, pancreatic carboxypeptidase, alkaline phosphatase, malic, lactic, and glutamic dehydrogenases, and tryptophan desmolase. It serves as an activator in many other systems, including arginase, carnosinase, oxalacetic decarboxylase, and several peptidases. It is vitally concerned in nucleic acid metabolism and cell division [CHESTERS, 1974], and plays undisclosed roles in the choroid of the eye, and in endocrine physiology.

A portion of the biological activity of zinc may be concerned with the maintenance of protein structure, in addition to its catalytic effect. Thus, in alcohol dehydrogenase, two of the four atoms of zinc are buried within the protein structure and maintain the structural unity of the enzyme, while two others are involved in maintaining the catalytic effect. A similar consideration may hold in regard to the zinc requirement for the alteration of expression of the genetic potential of cells during transformation from resting to preparative stages for DNA replication. The function of zinc in wound healing and in reproduction has long been suggested by observations of these processes in experimentally deficient animals, and a special role in DNA function seems assured.

Zinc deficiency has been observed in a variety of species, and although it has been observed naturally in pigs and in lambs, it is not a typical grazing disease. Marginal inadequacies are recognized in man, and may present a broad pediatric problem in children where the requirements are high and the consumption toward the lower range of the spectrum of dietary intake. Zinc deficiency symptoms include impaired growth and reproductive function, along with gross bone disorders and lesions of the skin and its appendages.

Selenium (atomic No. 34) is well established as a nutritional requirement in the higher animals and in bacteria, and many higher plant species are selenium accumulators. Early studies of selenium metabolism suggested a strong functional connection with vitamin E, and also with sulfur amino acid levels in the diet, but it appears that these observations were based upon metabolic interrelationships rather than any direct activity of the element. More recently, a number of enzymes have been studied in which selenium plays an active role: glutathione peroxidase; a cytochrome with a protein component similar to cytochrome b_5 and the heme of cytochrome *c;* formic dehydrogenase from *Escherichia coli* and *Clostridium thermoaceticum;* and protein A, a component of the glycine dehydrogenase complex [SCHWARTZ, 1974]. Interest in selenium originally attached to a number of diseases of grazing animals which were eventually attributable to chronic selenium toxicity, but subsequent studies have also demonstrated a remarkable list of domestic animal deficiency diseases. Among the latter are muscular dystrophy in many species, exudative diathesis in chicks, hepatosis dietetica in pigs, and periodontal disease in ewes. Selenium deficiency is apparently rare in humans. Requirements for selenium in domestic animals are highly dependent upon the form in which the element is administered, and the amounts of vitamin E in the diet.

Molybdenum (atomic No. 42) is essential in animals as a component of the metalloprotein xanthine oxidase, and in plants as a portion of the metallo-enzyme nitrate reductase. Deficiencies have been induced in rats, chicks, turkey poults, and lambs, but natural deficiencies are unrecognized in any animal species. Molybdenum poisoning is known in grazing animals and is treatable with copper salts. Conversely, chronic copper poisoning has been treated with molybdenum. These antagonisms are also dependent upon the levels of sulfate in the diet, since sulfate can either ease or intensify the symptoms of molybdenum toxicity, depending upon the copper status of the animal. Naturally occurring deficiencies of molybdenum are unknown in animals, but molybdenum treatment of soils in deficient areas has produced appreciable increases in pasturage and thus, indirectly, in livestock gains.

Tin (atomic No. 50) has been shown to enhance the growth rate of rats by nearly 60% when 1 ppm of the element is added to a tin-deficient diet [SCHWARTZ, 1971]. It has been pointed out that tin fulfills all of the requirements for an element to function biologically, and further, that its chemistry places it in the fourth group of elements with silicon and carbon, so that it can form truly covalent linkages. It is also suitable for functioning as an oxidation-reduction mediator, having a potential very close to that of the flavin enzymes. Unfortunately, little is presently known regarding the biological chemistry of this element.

Summary and Comment

A consideration of the chemical element requirements of animals and their biological functions in relationship to the distribution of elements in the periodic table indicates that 99% of the structure of animals is synthesized from 13 of the first 20 elements, and that eight of the next ten elements serve obligately in a biocatalytic capacity. These facts reflect the relative abundance of the elements, which varies inversely with atomic number, and geometric considerations which complicate the intrusion of larger atoms into biological molecules. Beyond the first 30 elements, selenium is assured of a role in animal biology. Molybdenum is known to function in but a single animal enzyme system, there is only limited information available on tin as a growth stimulant in rats, and iodine serves a highly specialized function in higher animals.

The only elements of the first 20 (excepting the inert gases) that are apparently nonessential to animals for maintenance of the fluid environment (including ground substance) and for structural purposes are lithium (atomic No. 3), beryllium (atomic No. 4), boron (atomic No. 5) and aluminum (atomic No. 13). Since trace element research technology has now progressed to an advanced state, it is unlikely that these will ever arise as requirements for structural purposes. Boron is well established as a plant requirement for cell maturation and differentiation, and it may be that traces of these very small atoms of elements 3, 4, and 5 may function in the stabilization of certain macromolecular structures. The insolubility of aluminum and titanium oxides possibly limited their primordial availability for biological processes. The occurrence of germanium (atomic No. 32) in the fourth group along with carbon, silicon, and tin, and its relative abundance in the lithosphere, suggest that this element is a likely candidate for inclusion on the list of required elements. The series from germanium to bromine (required by some invertebrates) would then be filled with the exception of arsenic, which has long been suspect as a trace requirement.

Accelerating human modification of the trace element composition of the biosphere in recent times has become a growing concern as it relates to the etiology of animal disease. Some kinds of elemental antagonisms, particularly among the trace metals, have existed since life forms first evolved. Increased concentrations of some of the elements not formerly distributed in like amounts in the biosphere are now appearing, and may mask, substitute for, or potentiate the effects of known or as yet undiscovered needed elements. More significantly, they may antagonize or inhibit normal biological trace element function, bringing about an entire new spectrum of chemical requirements to moderate novel conditioned deficiencies. Ecological contamination is even at present of sufficient magnitude to produce such situations.

References

Beerstecher, E., jr.: Petroleum microbiology (Elsevier, Amsterdam 1954).

Carlisle, E. M.: Silicon a possible factor in bone calcification. Science *167:* 279–280 (1970a).

Carlisle, E. M.: A relationship between silicon and calcium in bone formation. Fed. Proc. Fed. Am. Socs exp. Biol. *29:* 565–570 (1970b).

CARLISLE, E. M.: Silicon as an essential element. Fed. Proc. Fed. Am. Socs exp. Biol. *33:* 1758–1766 (1974).

CHESTERS, J. K.: Biochemical functions of zinc with emphasis on nucleic acid metabolism and cell division; in HOEKSTRA, SUTTIE, GANTHER and MERTZ Trace element metabolism in animals, vol. 2 (University Park Press, Baltimore 1974).

COTZIAS, G. C. and PAPAVASILIOU, P. S.: Primordial homeostasis in a mammal as shown by the control of manganese. Nature, Lond. *201:* 828–829 (1974).

FLORKIN, M.: Biochemical evolution (Academic Press, New York 1949).

FRIEDEN, E.: The ferrous to ferric cycles in iron metabolism. Nutr. Rev. *31:* 41–44 (1973).

FRIEDEN, E.: The biochemical evolution of the iron and copper proteins; in HOEKSTRA, SUTTIE, GANTHER and MERTZ Trace element metabolism in animals, vol. 2 (University Park Press, Baltimore 1974).

FÜRST, P.; JOSEPHSON, B.; MASCHIO, G., and VINNARS, E.: Nitrogen balance after intravenous and oral administration of ammonium salts to man. J. appl. Physiol. *26:* 13–22 (1969).

HOPKINS, L. L., jr.: Essentiality and function of vanadium; in HOEKSTRA, SUTTIE, GANTHER and MERTZ Trace element metabolism in animals, vol. 2 (University Park Press, Baltimore 1974).

HOPKINS, L. L., jr. and MOHR, H. E.: Vanadium as an essential nutrient. Fed. Proc. Fed. Am. Socs exp. Biol. *33:* 1773–1775 (1974).

KIES, C.; TOBIN, R. B.; FOX, H. M., and MEHLMAN, M. A.: Utilization of 1,3-butanediol and nonspecific nitrogen in human adults. J. Nutr. *103:* 1115–1163 (1973).

KIES, C.; WILLIAMS, E., and FOX, H. M.: Effect of nonspecific nitrogen intake on adequacy of cereal proteins for nitrogen retention in human adults. J. Nutr. *86:* 357–361 (1965).

LEACH, R. M., jr.: Biochemical role of manganese; in HOEKSTRA, SUTTIE, GANTHER and MERTZ Trace element metabolism in animals, vol. 2 (University Park Press, Baltimore 1974).

MATHEWS, M. B.: Molecular evolution of Connective tissue; in PERSON Biology of the mouth (Am. Association for the Advancement of Science, Washington 1968).

MERTZ, W.: Chromium occurrence and function in biological systems. Physiol. Rev. *49:* 163–175 (1969).

MERTZ, W.: Trace element research: recent developments and outlook; Chromium as a dietary essential for man; in HOEKSTRA, SUTTIE, GANTHER and MERTZ Trace element metabolism in animals, vol. 2 (University Park Press, Baltimore 1974).

MILNE, D. B. and SCHWARTZ, K.: Effect of different fluorine compounds on growth and bone fluoride levels in rats; in HOEKSTRA, SUTTIE, GANTHER and MERTZ Trace element metabolism in animals, vol. 2 (University Park Press, Baltimore 1974).

NIELSEN, F. H. and OLLERICH, D. A.: Nickel: a new essential trace element. Fed. Proc. Fed. Am. Socs exp. Biol. *33:* 1767–1772 (1974).

PEISACH, J.; AISEN, P., and BLUMBERG, W. E. (eds): The biochemistry of copper (Academic Press, New York 1966).

PERSON, P.: Biology of oral tissues and oral disease: Darwin and Quantum; in PERSON Biology of the mouth (Am. Association for the Advancement of Science, Washington 1968).

SCHROEDER, H. A.: The biological trace elements. J. chron. Dis. *18:* 217–218 (1965).

SCHWARTZ, K.: Tin as an essential growth factor for rats; in MERTZ and CORNATZER Newer trace elements in nutrition (Marcel Dekker, New York 1971).
SCHWARTZ, K.: Recent dietary trace element research, exemplified by tin, fluorine and silicon. Fed. Proc. Fed. Am. Socs exp. Biol. *33:* 1748–1757 (1974).
UNDERWOOD, W.: Trace elements in human and animal nutrition; 3rd ed. (Academic Press, New York 1971).
WILLIAMS, R. J. P.: Heavy metals in biological systems. Endeavor *26:* 96–100 (1967).
WILLIAMS, R. J. P. and WACKER, W. E. C.: Cation balance in biological systems. J. Am. med. Ass. *201:* 96–100 (1967).

Prof. E. BEERSTECHER, Department of Biochemistry, Dental Branch, University of Texas at Houston, PO Box 20068, *Houston, TX 77025* (USA)

Comp. Anim. Nutr., vol. 2, pp. 28–86 (Karger, Basel 1977)

Metabolism of the Major Mineral Elements (Ca, P, Mg, S and also Fe) in Relation to Requirements

D. P. CUTHBERTSON

University Department of Pathological Biochemistry, Royal Infirmary, Glasgow

I. Role of Major Minerals in Marine Animals

A. Introduction

Although osmotic pressures of external media are usually equal in marine invertebrates, the concentrations of individual ions are frequently dissimilar. The ionic compositions of body fluids throughout the animal kingdom possess a general pattern of resemblance [NICOL, 1969].

The ability to regulate ionically is a universal characteristic of marine animals, differing in magnitude in the different groups. In more primitive groups regulation is slight. There is some regulation of K^+ and, to a lesser extent, of Ca^{2+}; SO_4^{2-} is reduced slightly, except in *Aurelia* and *Phallusia,* where it is halved. Na^+, Mg^{2+} and Cl^- are in equilibrium with sea water. In contrast, the two active invertebrate groups, cephalopods and decapod crustaceans, show well marked regulation of all ions in the haemolymph. Calcium is maintained at higher levels than sea water in most species examined. The most striking feature is the great reduction in magnesium ranging down to 14% of sea water values. In *Homerus* sulphate is likewise reduced. A peculiarity is seen in decapod blood where the reduction of magnesium is correlated with accumulation of sodium. Hermit crabs *(Eupargurus)* are unusual among decapods in accumulating sulphate. The isopod *Ligia oceanica* likewise shows well-marked ionic regulation. Sodium, potassium, calcium and chloride are all more concentrated than in sea water, whereas magnesium and sulphate are much reduced.

In the lower marine vertebrates, magnesium is greatly reduced in the plasma with the increase of sodium, and sulphate is held at very low levels.

B. Phyletic Review of Primitive Groups in Equilibrium with Sea Water

The ionic composition of the mesogloea in *Aurelia* differs slightly from that of sea water, and is controlled by the bounding ectodermal and endodermal epithelia. Sulphate is eliminated with associated cations – a passive consequence of the reduction in sulphate. The body fluid in consequence has a lower specific gravity than sea water: the replacement of some 40% of that amount of sulphate occurring in sea water by chloride renders the animal naturally buoyant. This phenomenon is seen in other gelatinous planktonic animals, such as ctenophores, etc.

Sea anemones have calcium concentrations nearly equivalent to sea water, their body walls show two-way permeability to water and calcium, but exchange of the latter is rather slow.

The concentrations of ions in coelomic fluids of echinoderms are very similar to those in sea water. Potassium is regulated in all species, and magnesium in some instances.

There is selective absorption of ions by the body wall of polychaetes and others resulting in a urine low in potassium and rich in sulphate.

1. Decapod Crustaceans

Marine decapods generally possess high levels of sodium, potassium and calcium ions, and reduced levels of magnesium and sulphate ions. Lobsters are peculiar in having (1) low potassium values; (2) active absorption by the gills of sodium, potassium, calcium and chloride against a concentration gradient, and (3) inward diffusion of magnesium and sulphate along the concentration gradient – the differential gradient excretion by the antennary glands tends to lower blood magnesium and sulphate, and conserve sodium and potassium. The gills and integment of the lobster are relatively impermeable to magnesium and sulphate ions, which enter largely through the gut.

Examination of crustacean urine reveals differential excretion. Sodium, potassium and calcium are conserved, while magnesium and sulphate are eliminated.

2. Mollusca

In lamellibranchs and gastropods, calcium along with potassium is regulated, and in cephalopods regulation extends to all ions.

Examination of the fluid from the renal sac of cephalopods shows that resorption of potassium, calcium and magnesium ions, and secretion of sulphate ion take place in the formation of urine. As a result, levels of the former ions are raised, and levels of the latter lowered in the blood. As far as absorption is concerned, it would appear that potassium, magnesium, etc., are taken up against a concentration gradient, while sodium and sulphate enter along a diffusion gradient.

3. Tunicates

All ions except sodium are kept at different values from those of sea water: potassium is significantly raised in *Salpa,* and the divalent ions – calcium, magnesium and sulphate are reduced in tunicates, the whole burden of ionic regulation is borne by the external surfaces [ROBERTSON, 1949].

4. Fishes

Myxinoids have fluids slightly hyperosmotic to sea water although there is no regulatory mechanism for sodium and chloride, potassium and the divalent ions – calcium, magnesium and sulphate are secreted into the urine.

Marine elasmobranchs tend to be almost isosmotic with sea water owing to high internal concentrations of urea. Magnesium and sulphate ions are absorbed to only a slight extent, compared with potassium, calcium and chloride ions. Excess magnesium, sulphate and chloride are excreted by the kidney.

In teleosts, the ionic regulation is somewhat similar except that the blood is hypoosmotic. Calcium, and especially magnesium and sulphate, are concentrated in the intestinal fluid. Excess magnesium and sulphate are excreted extrarenally via the gills.

C. Major Minerals in Tissues and Skeletons

Iron is found in cytochrome and widely distributed in intracellular haemochromogen and in certain respiratory pigments (haemoglobin, chlorocruorin, haemerythrin). The radular teeth of chitons (Chitonidae) and limpets (Patellidae) contain large amounts of iron as Fe_2O_3 (54% of ash in *Patella vulgata)*. The iron content of sea water is low (0.002–0.02 mg/kg) and *Natilba* relies mainly on algal food supplies for iron. The iron content of certain worms, *Lineus* (nemertine) and *Nephthys* (polychacete) is high [JONES *et al.*, 1935].

Calcium is extensively utilized in skeletons of animals, and in calcareous tubes.

$CaCO_3$ occurs in coral skeletons and molluscan shells, and as $CaCO_3$

and $Ca_3 (PO_4)_2$ in the exoskeleton of decapod crustaceans. Magnesium is an important constituent of the skeleton of certain animals – Foraminifera, Alcyonaria, Echinodermata and Crustacea. Magnesium is very high in some gastropods (1.58% by weight in *Achidoris)*.

1. Eggs, Embryos and Larvae

In many egg cells, calcium and magnesium are present in colloidal combinations.

2. Skeletons of Marine Animals

Hard skeletons provide for support, protection and defence: endoskeletons protect soft and delicate tissues, e.g. the brain of cephalopods, cyclostomes and fishes. They further give internal support to the body wall and appendages, e.g. in cephalopods and fish, and form rigid supports and lever systems for the functioning of muscles in certain mollusca, in crustaceans and chordates. The exoskeletons of different animals have similar protective and functional roles and also provide barriers to the diffusion of water and solutes.

Calcium differs from the transition elements in having electrons in its inner orbits which are relatively stable. In consequence, it does not react with many covalent bonds in the way that a transition element, much as copper would, nor does it act as a potent enzyme inhibitor, and is not poisonous in physiological systems.

The majority of animal shells are calcareous, the carbonic acid being derived from endogenous carbon dioxide and it salts are frequently retained by animals and function as buffers. Carbonic anhydrase is formed at many sites of calcification.

The phosphate anion shares many of the properties associated with carbonates. It is of great importance to the metabolism of all cells and is retained in the body where it forms an important component in the buffer system particularly in relation to urine.

The insolubility of shells arises from the fact that both anions have a large size and are strongly charged. They are, in consequence, likely to lose electrons when close to multivalent cations so that there is a tendency towards covalency and a spreading out of the electron cloud. The result of this may be regarded as either weakening the attraction to water molecules or alternatively of making it more difficult to fit the salt into the water lattice.

The minerals of calcified tissues orginate from the physiological fluids by the formation of a crystalline solid phase.

KENNEDY *et al.* [1969] point out that shells of bivalves may be wholly aragonitic, or may contain both aragonite and calcite, in separate monomineralic layers. Shells are built up of several layers of distinct aggregations of calcium carbonate crystals and these aggregations are referred to as shell structures. Aragonite occurs as prismatic, nacreous, crossed, lamellar, complex cross-lamellar and homogenous structures. Calcite occurs as prismatic or foliated structures. Myostracal layers are always aragonite. The ligament and byssus when calcified are also aragonite. Calcite is found only in the outer layer of superfamilies belonging to the subclass Pteriomorphia with no exceptions. In a superfamily, the shell structure and mineralogy are very constant and these combinations have existed for millions of years.

The prime control on shell mineralogy is genetic. Environmental factors may modify the basic mineralogy/shell structure further within a superfamily. There is an inverse relationship between the percentage calcite in the shell and the mean temperature of the environment inhabited by the bivalve.

The species *Mytilus californianus,* which shows the biggest temperature effects, is peculiar amongst Mytilacea in having an inner as well as an outer calcite layer.

The control of periodicity of shell deposition is relatively unknown. Ecological factors, such as day/night cycle, tides and food intake may play a role. The giant clam *Tridaena gigas* was found to form approximately 285 layers in one year, suggesting daily layers for a greater part of the year [BONHAM, 1965].

D. Factors Influencing Requirements of Aquatic Animals

Little is known concerning the mineral requirements of aquatic animals, although much is known about their osmotic regulation and their chemical composition. Rates of accretion can be determined and periodicity established. The group of marine animals in which calcium is obviously very necessary are the molluscs for the shell, crustacea for the carapace and coral for the skeleton.

The exchange of ions from the environment across the gills and skin of fish complicates determination of their daily quantitative requirements. Calcium and phosphorus are absorbed directly from the environment: their actual ingestion is not necessary for absorption.

Sulphate ions may also be absorbed from the water, but both phosphorus and sulphur are obtained more effectively from the food. In their absorption from water, the uptake is proportional to the concentration of the anions in the water provided the levels do not affect the fish.

The USA National Research Council [1973, No. 11] point out that calcium absorption is nearly independent of the calcium concentration within the ranges of 5–500 ppm after a 24-hour acclimation period. Magnesium depresses calcium absorption. More calcium is stored in the skin and less in the skeletal tissue in the presence of copper and zinc.

Sea water contains considerable quantities of calcium (25 mEq/kg), magnesium (129 mEq/kg) and sulphate (68 mEq/kg), and as marine teleosts (true or bony fish) drink sea water it is probable that they have no dietary requirement for these minerals.

E. Marine Birds and Mammals

1. Marine Birds

It seems appropriate to consider here the food of marine birds which spend much of their lives far beyond sight of land. The adequacy of their intake of minerals is assured if the quantity of food can be achieved, for their food consists to a large extent of surface plankton, including crustacea, jelly fish, molluscs, as well as squid and small fish. Into this category fall petrels, shearwaters, fulmars and albatroses *(Tubinares)*.

Penguins eat fish and plankton; kittiwakes and swallow-tailed gulls eat fish, crustacea, molluscs, offal from fishing ships; tropical birds eat fish and squids.

2. Marine Mammals [for review, see NICOL, 1969]

Carnivorous marine mammals include whales, seals, sea otters and even bats. The toothed whales (Odontoceti) are hunters and exploit many forms of nekton (strong swimming animals, such as squid, fish and whales). The sperm whale *Physeter* feeds on fishes and especially cephalopods. Porpoises and dolphins are voracious eaters of small fish. The killer whales *(Orca)* eat whole porpoises, seals and kill walruses and large whales.

Some seals are planktonic feeders. Crab-eating seals *(Lobodon carcinophagus)* of the antarctic are selective feeders consuming krill. Fish are eaten by the eared seals *(Callorhinus ursinus)*. Elephant seals *(Mironuga leonina)* of South Georgia capture cephalopods and walruses *(Odobenus)* dive after bivalves. In antarctic waters, the large leopard seal eats fish and Cephalopods, and attacks penguins and other seals.

The sea otter *(Euhydra lutris)* eats hard-shelled invertebrates – bivalves, sea urchins, abalones, etc.

Piscivorous bats *(Noctilio, Pizonyx)* capture fish at the surface by using their hind legs for the purpose.

The mineral adequacy of the diets of these marine mammals is assured if they can capture sufficient food for their energy requirements.

II. Requirements of Fish

There is very little on the mineral requirements of fish apart from what is surmised from our knowledge of higher animals. It has been pointed out that many required minerals can be absorbed through gills and other body surfaces by fishes, but that soft waters which are poor in minerals necessitate either (1) fertilization of the water with the needed minerals or (2) suitable mineral supplementation of the fishes' diet. The second method is more practical when fish, such as trout, must be reared in flowing water.

All minerals essential to adequate nutrition of higher animals can be considered essential for fish until proved otherwise [ASHLEY, 1972]. Of the antianaemic minerals, iron heads the list except in molluscs and in certain arthropods. Some fish are so deficient in iron that when fed to mink (or to other fish) induce anaemia. Cooking with 16 mg organic-bound iron added per week prevents anaemia in mink [ASHLEY, 1972].

Calcium deficiency may not show up so obviously in fish since fish bones are less highly calcified than those of mammals. A rachitic condition has not been observed in fish.

There is no recorded case of magnesium deficiency occurring in fish, but it is an essential element for them.

Inorganic fertilizers in pond fish culture have been recently employed. Calcium and phosphorus play important roles in vegetable and plant growth, the former is particularly important in acid ponds and liming is a common practice.

Whereas on the carbon and nitrogen uptake of copepods there is a large literature, there is virtually nothing on minerals apart from observations on the plankton as a factor in nitrogen and phosphorus cycles in the sea, and since most mineral nutrients in the sea are in solution in the water they may be absorbed through the gills.

Studies by PHILLIPS *et al.* [1964], TEMPLETON and BROWN [1963], ICHICKAWA and OGURI [1961] and SMELOVA [1962] have shown that calcium, phosphorus and sulphate are taken directly from the water by trout. They have also demonstrated the absorption and distribution of these minerals from the food by and to the body tissues. Other workers have found similar results with other fish.

Modern dry diets permit trout to be grown, starting from swim-up stage, exclusively on dry feed. Nutritional requirements of salmon have also been investigated in a similar fashion, but seldom have the requirements been defined in terms of actual nutrients. For example, the dry pellet mixture of PHILLIPS *et al.* [1964] is as follows: bone meal, 5.0%; fish meal, 24.0%; dried skim milk, 3.5%; cotton seed meal, 5.0%; wheat middlings, 7.0%; brewers' yeast, 10.0%; cellulose flour, 20%; vitamin mixture 1.5%; A-D feeding oil, 30%. On such a diet and at a temperature of 13–15 °C, trout will grow to market size within a year. HEVESY *et al.* [1964] found that at 18 °C intramuscularly injected ^{59}Fe was incorporated into the blood corpuscles of tench to a maximum of 70% and at 5 °C only 4% was taken up by the erythrocytes. Effective artificial feeding of several warm water fishes is routine in several areas for carp, cat fish, eel, yellow tail, milk fish, cultivated crustaceans and tilapia. The natural food of carp is primarily from crustacea, worms, insect larvae, small molluscs, ingested.

Soluble inorganic salts may be absorbed involuntarily with other products of digestion e.g. by prawns: some may be lost by regurgitation. The former would be secreted back into the gut and released in solution through the anus – not in the faeces – as part of the osmoregulatory process [DALL, 1968, 1970].

One of the most recent positive studies is that of ANDREWS *et al.* [1973] on the effects of dietary calcium and phosphorus on growth, food conversion, bone ash and haematocrit levels of catfish. Channel catfish *(Ictolurus punctatus)* fingerlings were given diets containing different amounts and ratios of calcium (0.5–2.0%) to phosphorus (0.5–1.2%). The results of 2 studies indicated that the available phosphorus requirement of catfish is about 0.8% of the diet. In both experiments, gains were greater in fish on diets containing 1.5% calcium and were reduced when more was given. The rise in blood calcium could not be prevented by adjusting the Ca:P ratio by the addition of phosphorus to the diet. Phosphorus deficiency characterized by reduced growth, poorer feed efficiency and lower bone ash, and lower haematocrit occurred in fish given 0.5–0.6% of available phosphorus. The absence of a growth response to 0.4% added phosphorus in the form of calcium phytate suggested that catfish are not able to use phytin-P. Bone ash values from skull and vertebrae samples indicated that the calcium requirement for maximum bone mineralization may be higher than for optimum growth.

Bones of sea fish from the Bay of Bengal or from the Dacca rivers contained 7.5% phosphorus and 20.3% calcium in Carcharius spp.; 5.1% phosphorus and 11.5% calcium in *Stromateus niger; Hilsa ilsha* in fresh water, 7.2% phosphorus and 17.5% calcium; the river fish *Wallago attu,* 8.4% phosphorus and 19.4% calcium, and *Channa punctata,* 8.6% phosphorus and 20.6% calcium [BASHIRULLAH and ANAM, 1970].

It is recorded that growth of brown trout *Salmo trutta* (L) in Ireland is poor where calcareous drift has made the water alkaline [KENNEDY and FITZMAURICE, 1971]. This paper gives in considerable detail the food intake as revealed by examination of stomach contents. This reveals the very considerable variety of minute animals eaten. Cladocera in particular, was extensively cropped as well as molluscs, larvae, pupae, emerging chironomids [KENNEDY and FITZMAURICE, 1971]. Similar observations have also been made in the Sierra Nevada lakes of California (by ELLIOTT and JENKINS [1972] who recorded both surface and aquatic organisms eaten by the trout – again demonstrating the very large variety of minute organisms consumed, diptera larvae and pupae, coleoptera and trichoptera, zooplankton and bivalve molluscs. There seems to be a degree of specialization exhibited by inidvidual trout [BRYAN and LARKIN, 1972]. Many other fish have been similarly investigated, e.g. herring, cod, mackerel, hake, etc.)

III. Insects

Due to a dearth of accurate information, it is usual to provide for minerals in insect experimental diets by the inclusion of one or other of the popular mammalian salt mixtures. However, the fact that a need for calcium could not be demonstrated for *Drosophila* [SANG, 1956], and could therefore have been required in no more than trace amounts, indicates that such mixtures may be far from ideally balanced for insects. Locusts are very insensitive to both the overall amounts of minerals in the diet and the relative proportions of the radicals present. A simplified mixture of four salts comprising Na, Ca, K, Mg, CO_3, Cl, PO_4 and SO_4 ions was shown to allow normal amounts of locust larvae [DODD, 1961].

The salivary glands of the adult blowfly, *Calliphora erythrocephala* have unique properties. The secretion is regulated by 5-HT [BERRIDGE and PATEL, 1968]. Three major events have been recognized during the stimulation of fluid by 5-HT: (1) 5-HT recognizes and interacts with a specific cellular receptor; (2) the result of such a reaction is decoded into an increase in the concentration of two intracellular messengers, cyclic AMP and calcium, and (3) cyclic AMP and calcium are then responsible for stimulating the transport processes of the cell to produce an increase in fluid secretion.

Concerning phosphorus metabolism during the period of metamorphosis, the fly *Calliphora erythrocephala* has received considerable attention. There is a phosphate fraction which is difficult to hydrolyze and this may be a phosphorus-containing peptide or a complex-linked glucose amino phosphate, possibly some kind of a chitin precursor [AGRELL, 1953]. The chitin formation in endopterygotes occurs consistently in two steps, one for pupal and the other for imaginal chitin [AGRELL, 1953]. The level of immediately available energy in the pupa is not an uncomplicated reflection of the general energy production. The levels of other energy-rich di- and tri-phosphates, much as UTP, UDP, STP, GDP and CTP, do not vary along U-shaped curves but rather along N-shaped curves.

IV. Metabolism of Calcium and Phosphorus in Relation to Requirement

Calcium and phosphorus are present in bone and teeth of higher animals as members of the hydroxyapatite family, $Ca_{10} (PO_4)_6 (OH)_2$, and are also mainly present in the earth's crust in this form. By leaching of primary rocks on the earth's surface through geological time, living organisms have been supplied with a source for biological calcium phosphate formation. POSNER [1969] has given details of marine and land calcium phosphate cycles. The relative insolubility of calcium and phosphorus salts in nature probably accounts for the relatively low concentration of calcium in foods. Bone and teeth are specialized tissues characterized by the presence of cells with long branching processes (osteocytes) which occupy cavities (lacunae) and fine canals (canalicula) in a hard, dense matrix consisting of bundles of collagenous fibres in an amorphous ground cement impregnated with hydroxyapatite. In the actual formation of bone, distinctive large cells of mesenchymal origin (osteoblasts) become surrounded by bundles of collagenous fibres which are then cemented together to form spicules or plates of uncalcified bone matrix (osteoid). Minute crystals of calcium and phosphate complexes are deposited in the osteoid. The osteoblasts imprisoned in the matrix are converted into osteocytes.

Bone is not a static tissue, but is continually reconstructed or remodelled throughout life. Where bone is being removed, the matrix shows a sharply 'bitten' surface; the bites or 'howships' lacunae being occupied by large multi-nucleated cells termed osteoclasts. When this happens, the fibres, cement and calcium salts all disappear together.

Teeth consist of three different calcified tissues: enamel, dentin and cementum. The bulk of the tooth consists of dentin, the crown is covered by enamel and the roots by cementum. Enamel is the hardest tissue in the body and is formed by the enamel organ, a complicated structure in which the active cells are the ameloblasts. An organic matrix is first laid down and is gradually thickened by increments to its final width, becoming calcified at the same time, the increase being linear. During this process, the organic matrix is withdrawn, so that finally there is virtually none in mature enamel which is 95–97% inorganic matter [HODGE, 1938].

Calcium is involved in several functions within the contraction of striated muscle. It is important in the steady state permeability of the muscle membrane and in the changes in permeability that produce the action potential. The level of depolarization necessary to initate contraction appears dependent upon superficial calcium sites, and the intensity by the intracellular concentration of calcium ions. The duration of contraction is controlled by the rate of removal of intracellular ions by the sarcoplasmic reticulum. WINEGRAD [1969], who has summarized its action, also points out that the synthesis of energy-rich phosphate compounds may be coupled with excitation by the activity of calcium-sensitive enzymes which control the rate of glycolysis. Thus, concludes WINEGRAD, calcium plays a crucial role in the action potential, which can be likened to a switch, and in the excitation-contraction coupling mechanism acts like a rheostat.

A. Calcium and Phosphate Metabolism

Calcium and phosphate homeostasis is controlled by shifts of calcium and phosphate among five different compartments: (1) the extracellular fluids (ECF); (2) the intracellular pool, divided into several compartments; (3) the bone and bone fluids; (4) the intestinal lumen, and (5) the renal tubular fluid [BORLE, 1974]. Three hormones: parathyroid hormone (PTH), calcitonin (CT) and cholecalciferol (vitamin D_3 – now recognized as a hormone) regulate the transport between the compartments. Of these hormones, PTH plays the principal role in calcium homeostasis. It maintains the concentration of ionized calcium in the extracellular fluids at a constant level and immediately corrects any fall in plasma calcium. It also plays a role in bone metabolism, specifically in bone resorption. The effects of calcium on PTH secretion may be mediated by cyclic AMP: the inhibitory effect of calcium on PTH secretion rate could be due to the fact that its adenyl cyclase is uniquely sensitive and inhibited by extremely low concentrations of calcium [CARE and BATES, 1972; DUFRESNE and GITELMAN, 1972]. Calcitonin also stimulates PTH secretion. Since calcitonin may depress intracellular calcium, its action could be explained by a depression of adenyl cyclase leading to an increase intracellular cyclic AMP concentration and an increased PTH secretion.

Calcitonin may protect the organism from the hypercalcaemia which follows calcium absorption [MUNSON *et al.*, 1973]; it may prevent the rise in plasma phosphate which often accompanies hypercalcaemia [TALMAGE *et al.*, 1973]. It may act as a rapidly acting damping factor to suppress oscillations of plasma calcium above the physiological levels. PTH controls calcium on the physiological range and suppresses oscillations of calcium below the physiological level [ARNAUD *et al.*, 1970].

The formation of the new hormone 1,25-dihydroxycholecalciferol (1,25-$[OH]_2D_3$) in the kidney, under the control of a hydroxylase regulated by PTH, by calcium or by phosphate, constitutes a new endocrine feedback system which may ultimately account for the interdependence of PTH and vitamin D, also for the adaptation of intestinal calcium absorption, and for the pathology of some vitamin D-resistant diseases.

B. Calcium and Phosphate Absorption

In the intestine, calcium absorption by the cells involves at least two steps; calcium uptake on the mucosal side and its corollary calcium efflux in the seroid side. At least one of these steps is an active process – probably the latter. In this process of absorption, several factors are involved: a calcium-binding protein, an exchange of calcium for sodium, calcium transport into mitochondria, and the intervention of the enzymes alkaline phosphatase and calcium-sensitive ATPase [BORLE, 1974]. The accumulation of calcium in mitochondria may be energy-dependent.

Electron microscopy of the small intestine reveals that in vitamin D deficiency there is virtually complete disappearances of electron-dense calcium granules from the mitochondria, whereas after D administration mitochondria contain numerous and dense calcium granules [SAMPSON *et al.*, 1970].

The most recent information suggests that phosphate is not required for calcium absorption and that phosphate absorption is independent of calcium transport. There is general agreement that a low-calcium diet results in an increased calcium absorption [BORLE, 1974]. Such a diet tends to depress the plasma calcium, thereby increasing PTH secretion which favours the conversion of 25-(OH)D_3 to 1,25-$(OH)_2D_3$ in the kidney [BOYLE *et al.*, 1971; GARABEDIAN *et al.*, 1972] which would result in increased calcium absorption.

Calcium absorption is unaffected by dietary phosphate, but plasma calcium does vary with the Ca:PO_4 ratio of the diet: with a high Ca:PO_4 ratio the plasma calcium tends to be elevated if PTH is indirectly shown to be decreased; with a low Ca:PO_4 ratio, plasma calcium is normal and PTH activity is presumed to be increased.

A calcium-sensitive ATPase, located on the brush border, may play a fundamental role in the cellular uptake of calcium [BORLE, 1974]; it is controlled by vitamin D. A calcium-binding protein has been found in many different species and in different tissues.

Renal function would seem to play a major role in determining serum calcium; however, in most species only one tenth of the extracellular calcium pool is excreted daily by the kidney, but in other species, for example the golden hamster, the daily calcium excretion is 12 times larger than the extracellular pool [BIDDULPH *et al.*, 1970].

It is still not settled that calcium homeostasis is controlled by cellular transport of calcium ions without involving bone formation or resorption [for a full discussion, see BORLE, 1974].

Calcitonin has been demonstrated in the ultimobranchial glands of amphibia and reptilia. In aves, the thyroid, parathyroid and ultimobranchial glands are located along the carotid artery on each side.

There is now general agreement that calcitonin is produced by the parafollicular or C cells which are formed in the thyroid and internal parathyroid (parathyroid IV).

In man, the concentration of calcitonin is low in the thyroids by comparison with rat, porcine, and bovine thyroids.

It would appear that calcium concentration in plasma is the controlled signal which is regulated round a reference value of approximately 10 mg% in normal animals. The controlling system is bone, and the controlling signal is a balance between bone anabolism (Vo+) and catabolism (Vo–). But the parathyroids and ultimobranchial cells contribute to precise regulation. Their actions are analogous to glucagon and insulin in the regulation of blood glucose.

Diets which are adequate in content of calcium are likely to contain more than the necessary amount of phosphorus so far as man is concerned. There is no evidence that a dietary deficiency is likely to occur. The ratio of CA:P in human milk is 2:1 whereas that advised for young infants is about 2.5–1.5:1. How far this is simply an adjustment to fit in with a diet where cows' milk preponderates with a ratio of 1.2:1 is uncertain, but probably is the conditioning factor. For adult man the phosphorus allowance should be approximately 1:1.5 (Ca:P). When computation of the total phosphorus intake in diets consisting largely of cereals is made, allowance has to be made for the relative poor absorption of phytin-P which is poorly absorbed unless the supply of vitamin D is adequate. Deficiency of phosphorus during pregnancy in humans is unusual.

HEGSTED [1971], reviewing the literature, considers that there seems to be very little evidence that the calcium:phosphorus ratio in the diet is of any practical importance; or, if it is of importance for good nutrition, it can be obtained on diets varying widely in the calcium:phosphorus ratio. It may be that in many investigations there have been coincident factors affecting the situation, namely low calcium intake with low supply of vitamin D plus possibly a genetic factor.

HEGSTED also points out that there is a similar voluminous literature on the effects of the special phosphate compound phytate, on the availability of calcium. Large amounts of phytate have been used therapeutically to inhibit calcium absorption, yet the practical significance of phytate continues to elude us. The adaptation of man and animals to diets high in phytate is still a mystery.

Diets high in phosphate inhibit utilization of calcium, magnesium, iron, zinc and probably many other minerals. The same is true with phytate. HEGSTED emphasizes that this reference to the solubility of phosphates has rarely, if ever, been demonstrated to be the 'cause' of the poor utilization. Consideration of the relative amounts of iron and phosphate in the diet make it all the more surprising that any iron is absorbed. It must be remembered that iron and zinc and many other trace metals are not present as free ions, but are co-ordinated with a ligand.

An important observation on senescent mice has been made by KRISHNARAO and DRAPER [1972]. They found that on a diet containing 1.2% calcium and 1.2% phosphorus, the femora showed a lower breaking strength and a reduced content of ash, calcium and phosphorus. The accelerated rate of bone resorption was arrested by reducing the phosphorus content of the diet from 1.2 to 0.6%. The exacerbating effect of the relatively high-phosphorus diet on the development of senile osteoporosis was confirmed by estimations of ^{45}Ca exchange. The effect was attributed to a mild secondary hyperparathyroidism resulting from a depression of serum calcium concentration caused by a flux of excess phosphorus through the blood.

The mineral contents of fetuses of pregnant rats at the 22nd day of gestation appeared to be unaffected by the level of calcium in the diet or by the amount of calcium retained by the mothers during pregnancy [SEKI, 1971].

DENIS *et al.* [1973] conclude that in young rats, dietary protein independently of parathyroid hormone or calcitonin determines the magnitude of the physiological and histological changes observed in growing rats given diets inadequate in phosphorus and calcium.

In the rabbit, it has been shown that when the diet is arranged so that a high-calcium, low-phosphorus diet alternates with a low-calcium, high-phosphorus diet, the retention of calcium dropped from 69 to 28% and phosphorus from 54 to 36% of that absorbed: calcium in the urine rose from 21 to 50% of that ingested and phosphorus from 28 to 44% [GUÉGUEN and TRUDELLE, 1972].

C. Calcium and Bone

Bone is both a crystalline solid and a living system able to adapt its structure to chemical and mechanical stress. It provides an enormous calcium reservoir for maintenance of the regulation of calcium ions in the blood. RICHELLE and ONKELINX [1969] conclude that it is the mineral phase of bone that is deposited first and as an amorphous material; as described by EANES *et al.* [1966]. This amorphous material gives rise to an initial crystalline precipitate. In turn, the initial crystalline precipitate is transformed subsequently into a final crystalline precipitate. Once the precipitate is formed, crystallization goes on at the expense of both of the amorphous precipitate and of the liquid phase. The apparent crystal size does not vary much with time and the increase in mineral load is mainly due to an increase in the number of crystals present. The initial crystalline precipitate is an apatite with a composition close to that of octocalcium phosphate (8 $Ca/6PO_4$). The final crystalline precipitate is an apatite similar to hydroxyapatite, $Ca_{10}(PO_4)_6(OH)_2$. Carbonate is present in at least two chemical forms and/or environments, the proportions of which vary with time.

Changes in the calcium content of a bone volume element which does not include Haversian canals or blood vessels, can occur only through a change in either its volume or its calcium density (mass of calcium per unit volume). The processes which affect bone volume are apposition and resorption; the processes that affect density are augmentation and diminution. The difference between the last two is the rate of secondary mineralization. When osteocytic osteolysis occurs, it is described by a diminution rate which exceeds the augmentation rate. However, according to MARSHALL [1969], calculation shows that osteolysis by erosion of the walls of the lacunae probably does not make a significant contribution rate to the diminution rate of normal bone. Calcium kinetics measure not the accretion rate but the *addition* rate, the sum of the *apposition* and *augmentation* rates. A calcium balance continued with calcium kinetics measures not the resorption rate but the removal rate: the removal rate is the sum of *resorption* and *diminution* rates in this sense resorption is restricted to the action of osteoclasts.

The short-term calcium appears to be located at bone surfaces. In the fully adult dog, augmentation diminution represents more than 80% of the long-term rate of calcium transfer, while apposition resorption less than 20%. In normal adult man, MARSHALL [1969] considers that augmentation diminution represents more than half the long-term rate of calcium transfer, apposition resorption representing less than half.

Radial diffusion within the cylindrical territory surrounding cannaliculi is presented as the probable mechanism for augmentation diminution and the diffuse component. This provides the explanation for the power function retention of alkaline earth radioisotopes.

Although apposition and resorption *are* more active during the period of rapid growth both processes continue, although at a reduced rate, in the adult [BAUER *et al.,* 1961].

In the case of wheat flour, the phytate content increases as the extraction rate rises. In the United Kingdom during World War II, it was decided to add chalk to the higher extraction then being used, since calcium absorption appeared to be likely to be adversely affected. However, it would appear that the fear of a slow decalcification of the population was exaggerated and WALKER *et al.* [1948] found in long-term experiments on human subjects who ate large amounts of high extraction flour that after a time they adapted to this interference with calcium absorption. Further, it had not been realized that phytic acid is hydrolyzed by phytase in the flour during baking and also that hydrolysis occurs in the human intestine. Furthermore, if adequate vitamin D is present, phytate-P is as available as inorganic phosphorus [BOUTWELL *et al.,* 1946; SPITZER *et al.,* 1948]. The long history of our forebears eating virtually 100% extraction flour without obvious harm suggests that phytates do not constitute a hazzard in western countries unless there is a deficiency of vitamin D coupled with insufficient calcium, leading to an imbalance in the Ca:P ratio, such as has occurred recently amongst chuppatie-eating Pakistani immigrants to the UK.

Recommendations of requirements for calcium and phosphorus imply that correct amounts of vitamin D are also being supplied. Low levels of dietary vitamin D or diminished synthesis in skin in such latitudes as Birmingham and Glasgow plus possibly a low intake of calcium plus some degree of resistance to calciferol may be responsible in some situations for the appearance of clinical rickets [SWAN and COOKE, 1971; WILLS *et al.*, 1972].

High phytate intakes are also held responsible for the disturbances of zinc as well as calcium metabolism prevalent among Iranian villagers [REINHOLD *et al.*, 1973]. The phytate-rich flat bread *tanok* (lavosh or shepherd's bread) is the major staple of the rural diet. *Tanok* resembles the chuppaties in its content of phytate.

Lack of exposure to sunshine has also been proposed as a factor in the aetiology of osteomalacia in the Beduin women of the Negev. These Beduin women live in brownish black tents and are practically completely covered in black clothes.

It is reported that injection of 75–100 μg (3,000–4,000 IU) of vitamin D/day appears to be associated with the so-called idiopathic hypercalcaemia found in infants. It is still uncertain if such massive doses of vitamin D every 6 months can be safely administered.

In animal husbandry, most authors express the requirement for pigs and poultry as a percentage of the air-dried diet, rather than as a daily requirement, but if the criterion is degree of bone mineralization the requirement is considerably higher. Adequate bone ash, calcium and phosphorus levels are achieved for growing pigs with a calcium concentration of 0.8% of the dry matter. But if it is desired to use high-energy diets of the sow's milk substitution type with a feed conversion efficiency of about 1:1, it has been suggested that the calcium level be increased to approximately 1%.

V. Metabolism of Magnesium in Relation to Requirements

About half the total body magnesium is in bone and accounts for 0.5–0.7% of bone ash; there is very little variation from one species to the next [MORGULIS, 1931]. Of the non-osseus tissues, liver and striated muscle contain the largest quantities, about 20 mEq/litre. The kidney and brain contain 17 and 13 mEq/litre, respectively, while the magnesium content of red blood cells is about 6 mEq/litre [SHOHL, 1939].

The magnesium in serum varies from species to species. In man, the normal mean varies from 1.8 to 2.1 mEq/litre and there is virtually no difference between infants and children and adults. In the turtle, it is high, namely 3.2, and in the carp and mouse low, 1.0 and 1.1 mEq/litre, respectively.

The protein-bound magnesium in serum constitutes some 25% of the total [WACKER and VALLEE, 1974]. All warm-blooded animals have almost the same percentage of magnesium-bound in protein, even though the protein varies from species to species. The percentage in cold-blooded species is generally higher [OGASAWARA, 1953]. Strangely, the concentration in the cerebrospinal fluid is higher than in the serum, so it is not an ultrafiltrate of plasma. Cows' milk contains about 10 mEq/litre [SMITH, 1957].

Magnesium takes part in a very large number of enzyme systems: most are in the form of metal-enzyme complexes although carboxylase is more nearly a metalloenzyme in its properties.

Magnesium activates the numerous and important enzymes which split and transfer phosphate groups, among them the phosphatases and the many enzymes concerned in the reaction involving adenosine triphosphate (ATP). In addition to thiamine pyrophosphate, magnesium is a cofactor in decarboxylation. Certain peptidases also need magnesium for their activity. The alkaline and acid phosphatases are activated by magnesium.

Magnesium is required also as a co-factor in oxidation phosphorylation. A need for this element in association with protein biosynthesis may be one of its most fundamental actions. The intracellular concentration of magnesium is about 10 mM in most tissues and remains relatively constant during a magnesium deficiency. The microsomal fraction usually contains both a higher concentration and a greater proportion of the total than other organelles and is likely bound to nucleic acids and proteins [HEATON, 1973].

The observations of ULLREY *et al.* [1962] have shown that very little of the magnesium of diets containing less than 0.04% magnesium is excreted by young pigs, suggesting high availability and low endogenous loss. If an average availability of 80% is assumed to hold for pigs of all weights, the daily requirement can be calculated. For live weights of 5, 25, 45 and 90 kg the daily requirements are 0.15, 0.29, 0.40 and 0.43 g, respectively [UK Agricultural Research Council, 1967]. The endogenous loss of this element from a 5-week-old pig is about 28 mg Mg/week.

COOK [1973], in a comparison of availability of magnesium salts, found that magnesium carbonate was the most available form as based on percentage absorption, percentage retention and femur magnesium concentration. Availability of magnesium as the chloride was nearly equivalent to that of magnesium carbonate. Although absorption and retention of magnesium provided as the oxidase, phosphate, sulphate and silicate salts were lower than those of carbonate or chloride, all the salts were nearly equivalent in their ability to support growth, plasma and kidney magnesium.

A. Magnesium Deficiency

In weaning rats on a diet containing 1.18 ppm Mg, vasodilation manifests itself by erythema, and hyperaemia appears within 3–5 days followed by pallor and cyanosis 12 days later. At 3–5 days there is also increasing neuromuscular hyperirritability, culminating in generalized seizures after the 15th day of experiment.

Calves raised exclusively on whole milk showed a condition attributed to magnesium deficiency. They failed to grow and developed skin lesions and neuromuscular hyperirritability, tetany and eventually death. Symptoms did not occur until the serum concentration had fallen from a normal of 2.2–2.4 mg/100 ml to below 0.7 mg/100 ml. The concentrations of calcium and phosphorus in the serum did not change. Thrombosis and perivascular necrosis of small blood vessels in many different tissues was the prominent histological lesion, similar to that in the rat [LOWENHAUPT *et al.*, 1950]. Although contents of calcium and magnesium of soft tissue were not changed significantly, both were lowered in bone; the magnesium content being decreased [BLAXTER *et al.*, 1954a, b]. Tetany did not occur until the animals had become depleted of 30% of their total magnesium. 15 mg Mg/100 g diet prevented the disease. For the maintenance of a level of 2 mg/100 ml serum throughout lactation, lactating beef cows required 22 g dietary intake of magnesium [O'KELLEY and FORTENOT, 1969].

1. Interrelationship between Body Temperature and Magnesium Concentration in the Blood

The serum magnesium concentration of hibernating species is increased during this stage [LYMAN and CHATFIELD, 1955; RIEDESEL and FOLK, 1956]. On lowering the body temperature of non-hibernating mammalian species and cold-blooded animals the serum magnesium increases [Platner, 1950; Platner and Hosko, 1953; Axelrod and Bass, 1956]. Further, the intravenous injection of magnesium induces a fall in the rectal temperature of dogs regardless of the existing ambient temperature [Haegy and Burton, 1948]. The reasons for this change are not understood.

The metabolic rate of magnesium-deficient rats is 125% higher than that of control animals kept at the same weight and which need only 8.3% of the total calories required by the deficient animals to maintain this weight [Kleiber *et al.*, 1941].

The average US diet contains about 120 mg magnesium / 1,000 kcal [USNRC, 1974]: typical british diets range from 150 to 450 mg [UK Dept. Hlth Social Security, 1969]. Magnesium is a constituent of chlorophyll. Cereals and vegetables between them normally contribute more than two thirds of the daily magnesium intake. A dietary deficiency is very unlikely.

Common soluble salts of magnesium are readily absorbed from the small intestine. There is no known factor like vitamin D which affects the absorption of magnesium as does vitamin D in respect of calcium.

B. Excretion

In man, about one third of that ingested is excreted in the urine. Magnesium injected intravenously is excreted quantitatively in the urine. It is filtered by the glomerulus and reabsorbed by the tubules, although the amount of the reabsorbed fraction is not known [Heller *et al.*, 1953].

The human requirements for magnesium during pregnancy and lactation have received but little attention.

Various studies have been made to determine whether magnesium depletion is a complication in infant malnutrition, particularly where there is deficiency of protein and energy. The studies of Caddell and Goddard [1967] demonstrated that in Nigerian children with severe malnutrition, muscle magnesium was low and most children hypomagnesaemic, and that there were fairly specific electrocardiographic changes associated with magnesium deficiency. However, Rosen *et al.* [1970] in a study of 100 malnourished native African children, hospitalized for their condition, did not confirm this. By random selection, half of the children in both groups had been having diarrhoea. In the unsupplemented group, plasma magnesium tended to fall by the 3rd day except in those who had initially low levels. But by the 10th day, even in the absence of supplemental magnesium, plasma levels had increased. There was no difference between the treated and untreated even in those who were initially hypomagnesaemic.

There was also no difference in mortality even in those initially hypomagnesaemic and no neurological changes or correlation between electrocardiographic, such as CADDEL and GODDARD [1967] had found. The authors suggest that the latter's findings may have been based on children who were more severely magnesium-deficient, as their staple diet was cassava.

Malnourished children are never as severely depleted of magnesium as they may be of potassium. This may be because muscle is the main reservoir of potassium, whereas bone subserves that function for magnesium. It may be that other nutrient deficiences may be operating rather than magnesium in such field studies.

In chronic alcoholics, diminished levels of muscle magnesium have been found, but no abnormal skeletal values have been noted. On repletion, increments in muscle magnesium occur [LIM and JACOB, 1972].

A dietary deficiency is unlikely to occur in man except during weight reduction in obese subjects and in chronic diarrhoea as in kwashiorkor.

VI. Metabolism of Sulphur other than as Sulphur-Containing Amino Acids

The foods ingested by animals contain many sulphur-containing compounds – mainly organic. Inorganic sulphur compounds comprise but a small fraction of the whole. Of the organic, methionine and cystine contribute the bulk of the sulphur and, as indicated in the beginning of this chapter, it is not our purpose to deal with their metabolism here as they are dealt with in the chapter dealing with protein and amino acid metabolism: of the remainder, the importance of thiamin, biotin, lipoic acid, coenzyme A, glutathione, ergothionine, thiolhistidine, taurine, sulpholipids, sulphated polysaccharides, djenkolic acid, and various thiocyanates and isothiocyanates of plant origin, and possibly other sulphur compounds has been recognized. DZIEWIATKOWSKI [1962] has provided a useful review of these compounds in metabolism. Knowledge of the direct oxidation of methionine is lacking, but it has long been known that the ingestion of this amino acid gives rise to increased urinary excretion of sulphate, taurine and thiosulphate. But these end-products may be the result of oxidation of molecules derived from methionine rather than methionine itself. Homocystine and cystine are the most probable derivatives.

A. Sulphated Compounds

a) Sulphated Polysaccharides

Such compounds are found widely distributed throughout the animal kingdom. The heparins aid in the maintenance of the blood's fluidity and chondroitin sulphates are involved in ossification mechanisms and in the hydration of tissues and the mucoitin sulphates lubricate and protect the gastro-intestinal tract [DZIEWIATKOWSKI, 1962].

b) Chondroitin Sulphates

The chondroitin sulphates probably exist in the tissues in combination with protein. Chondroitin sulpahte accounts for about 70% of chondron-mucoprotein; the rest is a protein which differs from collagen. Their role in ossification is far from clear. It is probable that growth of membranous bones and expansion in breadth of the diaphyses, as well as their remodeling, requires a prior production of chondroitin sulphates [DZIEWIATKOWSKI, 1962]. Similar mechanisms are probably involved in the development of epiphyseal ossification centres. Deficiencies of vitamins A and C, of growth hormone, and of thyroxine impair the synthesis of chondroitin sulphates. These sulphates play an important part in wound and fracture healing: they appear in the early reorganization of the tissues traumatized [KODICEK and LOEWI, 1955; UPTON and ODELL, 1956; LOEWI and KENT, 1957].

c) Mucoitin Sulphates

These are secreted by cells occurring along the entire length of the digestive tract. They are regarded as lubricants and anti-irritants of the intestinal mucosa.

B. Iron-Sulphur Proteins

Iron-sulphur proteins are defined by ORME-JOHNSTON [1973] as those in which iron is bound via a sulphur-containing ligand. This excludes haemo-proteins with axial sulphur ligands unless they also contain sulphur-liganded non-haemic iron atoms. Xanthine oxidase and related iron sulphur flavin enzymes are examples of the ubiquity of iron-sulphur proteins as components in biological oxidation systems.

C. Heparins

These anticoagulants of blood are widely distributed throughout the animal kingdom; their concentration in a given tissue bears a close relationship to the number of mast cells in that tissue [HOLMGREN and WILANDER, 1937]. These compounds are probably stored in the cytoplasm.

At least two forms have been found, α and β. The α form has a higher sulphur content. The β form is very likely a form of chondroitin sulphate (β form). The latter has only about $^1/_{22}$ the anticoagulant activity of the α form, but in systems with a low thrombin concentration, the β form is the more active antithrombic substance [DORFMAN, 1958].

Other types of sulphated polysaccharides have been found.

D. Thiocyanates and Isothiocyanates

The cyanide radical, whether administered in the inorganic form or produced in the body by hydrolysis of organic cyanides (nitrites), is rapidly and largely converted into the non-toxic thiocyanate radical.

Thiocyanate, which is normally found in the animal body, is derived not only from cyanide and thiosulphate by the enzyme rhodanase, but also from the free thiocyanates and isothiocyanate esters present in food.

Large doses of thiocyanate are effective in relieving high blood pressure. In addition, it is weakly goitrogenic in that it blocks the concentration of iodide by the thyroid gland, probably by interfering with the reaction whereby iodide is oxidized to iodine. Iodide is effective in counteracting the goitrogenic potential of thiocyanate.

Goitrogenic compounds occur in normal diets. A number occur in vegetables like cabbage, turnips, rape and various fruits. Most contain sulphur, for example, 1,5-vinyl-2-thiooxazolidone and 2-propenyl isothiocyanate. Unlike thiocyanate, some of them resemble thiouracil in their action in blocking the utilization of iodine in formation of diiodotyrosine [for review, consult DZIEWIATKOWSKI, 1962].

E. Thiamin (Vitamin B_1)

Vitamin B_1 is found in the tissues of animals almost entirely as the pyrophosphate (cocarboxylase) functioning as an enzyme in the decarboxylation of ketoacids, but it will not be discussed further here.

F. Sulphide Sulphur

The intact mammalian animal can handle large amounts of hydrogen sulphide. It is oxidased to thiosulphate. Ferritin may act as a catalyst for the oxidation of sulphide produced in the intestinal tract. This appears also to be the fate of ingested powdered sulphur.

G. Sulphate Sulphur

Sulphate is present in bone not only as sulphated polysaccharide, but also in inorganic form [DZIEWIATKOWSKI, 1962].

Sulphate for the conjugation of phenols can be both endogenous [REED *et al.*, 1949] and exogenous DZIEWIATKOWSKI [1962]. Sulphate sulphur is converted into taurine sulphur in embryonated eggs [MACHLIN *et al.*, 1955; LOWE and ROBERTS, 1955].

VII. Metabolism of Iron in Relation to Requirements

Iron is the fourth most abundant element in the earth's crust, yet iron deficiency has ranked second only to protein malnutrition in the number of people affected. It causes a hypochromic microcytic anaemia. The iron content of a 70-kg man ranges from 4 to 5 g and is distributed as follows: 70.5% in haemoglobin, 3.2% in myoglobin, 26% in storage compounds (ferritin and haemosiderin), 0.1% in transport form (transferrin), 0.1% in the cytochromes and 0.1% in catalase. Free ionic iron occurs only in negligible amounts: the large bulk of it is in the form of complexes.

A. Haemoglobin

Iron is found in the blood of all vertebrates, with the exception of certain pelagic transparent fish larvae *(Leptocephalus)*, and certain antarctic teleosts of the family Chaenichthyidae [NICOL, 1969]. It is the dominant respiratory pigment in annelids. Elsewhere, it is distributed in isolated instances among the hemichordates. pogonophores, holothurians, phoronids, arthropods, lamellibranchs and nemertines [HETZEL, 1963; RUUD, 1954; SOUTHWARD, 1963].

In animals with closed respiratory systems, haemoglobin occurs in erythrocytes or dissolved in the plasma. In certain terebellids *(Terebella* and *Travisia),* haemoglobin is in solution in the blood and the coelomic fluid is provided with erythrocytes. Amongst invertebrates with closed circulatory systems, there is a general tendency for haemoglobin to be dissolved in the plasma. As muscle haemoglobin (myoglobin) it also occurs in heart muscle of many fish and gastropods. It is also found in the nervous system of worms *(Urechis, Aphrodite,* nemertines) and lamellibranchs *(Tivela).*

The oxygen-transporting function of haemoglobin is due to haem, which loosely combines oxygen in the proportion of 1 mol O_2 to 1 atom of iron (oxygenation). More than one haem unit is combined with globin, the exact number varying with the species.

Iron compounds that facilitate electron transport include the cytochromes which are generally associated with oxidative phosphorylation in mitochondria, NAD-linked dehydrogenases and flavoproteins. Peroxidases and catalases are iron-containing oxidative enzymes [O'DELL and CAMPBELL, 1971].

B. Chlorocruorin

The prosthetic group of this greenish respiratory pigment is also a haem, but contains a different porphyrin form haemoglobin.

The affinity is on the basis of 1 mol O_2 to 1 atom of iron. It is confined to certain families of polychaetes in which it is dissolved in the plasma, and it may coexist with haemoglobin in the same species.

C. Haemoerythrin

It differs from haemoglobin in the absence of a metallo-pophyrin group. It is found in invertebrates, e.g. the polychaete *(Magelona)* and sipunculoids and priapuloids and in the brachipod *Lingula.* It is always enclosed in corpuscles. It combines with oxygen in the ratio $3Fe/O_2$: the coloured prosthetic group is not a porphyrin.

Haemocyanin (copper-protein compound) and haemovanadin (vanadium chromogen) are found replacing haemoglobin and sometimes coexisting with it. The former is the most important respiratory pigment of invertebrates. The latter is found particularly among ascidians.

D. Metabolism

1. Absorption

While the absorption of iron can take place from the stomach and from any place in the intestinal tract, it is principally absorbed from the duodenum and the ferrous form is preferentially absorbed over the ferric form. Ascorbic acid, which has the capacity to reduce iron to the ferrous state, thereby increases absorption [SANFORD, 1960]. Dietary factors which cause formation of insoluble complexes, much as phosphates or phytates, cause a reduction in iron absorption. However, absorption of haemoglobin iron is not decreased by phytate [BOTHWELL and FINCH, 1962].

Absorption of iron is influenced by the nutritive state of the subject, the form of iron in the diet and the nature of the other constituents. Normal subjects retain 5–10% and iron-deficient individuals 10–20% of the iron in the food consumed. While the normal subject should absorb 0.6–1.5 mg iron on an intake of 12–15 mg, the iron-deficient subject would absorb 2–3 mg/day. Increase in iron absorption has been observed in the course of iron deficiency anaemia during the latter half of pregnancy and certain stages of haemochromatosis and in general under conditions which produce increased haemochromatosis [O'DELL and CAMPBELL, 1971].

There is essentially no excretion of iron into the mucosal cells of the intestine. From these cells it passes into the blood stream directly without entering the lymphatic system. Iron absorption may be regulated by the level of unbound plasma transferrin [CONRAD and CROSBY, 1962] have proposed that absorption is controlled by the concentration of iron in the mucosa, which in turn depends on the amount of 'messenger iron' entering these cells from the plasma. It may represent the transferrin.

2. Transport

Transport is by way of the plasma and most of the iron leaving the plasma is taken up by the bone marrow and by the bone marrow for haemoglobulin synthesis. The total amount of iron in the circulating plasma volume of normal subjects at anyone time is 3–4 mg and the plasma undergoes a diurnal variation decreasing by as much as 10–30% from morning to evening.

Although 0.5–2 μg Fe/100 ml is circulating in the blood as haemoglobin, by far the greatest amount is complexed to the specific iron-binding, β_1-globulin, transferrin. This prevents binding by chelating agents such as phosphate or citrate. Normal human plasma contains 0.24–0.28 g transferrin/100 ml. About 30% of the transferrin is normally saturated with iron;

the remaining 70% represents a latent or unbound reserve [RATH and FINCH, 1949]. At saturation levels above 60%, much of the iron is deposited in the liver. Ferritin is the storage form with iron moving in the reduced form, but stored again as ferric in ferritin [MAZUR *et al.*, 1960].

At levels of transferrin saturation between 30 and 60% transfer of iron to reticulocytes occurs primarily in the bone marrow. It is transferred selectively from transferrin to red cell precursors [JANDL and KATZ, 1963; KATZ and JANDL, 1964].

3. Storage

In normal man, storage of iron in liver, bone marrow, spleen and other tissues constitutes about 25% of the total.

These iron stores, primarily ferritin and haemosiderin, provide a reserve for metabolically functional compounds, such as haemoglobin, myoglobin and the iron-containing enzymes.

4. Excretion

Iron is not significantly excreted in the urine of man and the iron in the faeces represents primarily unabsorbed iron. Rats secrete significantly more (10% of the intake) in the urine. Animals tend to hold on tenaciously to iron liberated from lipid red cells. There is a small loss of iron from the gastro-intestinal tract because of the rapid turnover of the mucosa, and biliary losses are small. Dermal losses occur and may contribute to iron deficiency in hot humid regions [FOY and KONDI, 1957].

E. Iron Requirements

The requirements for iron have recently been assessed [CUTHBERTSON, 1973; US National Research Council, 1974], as well as some of the factors affecting its absorption (table I).

In 1970, the Food and Nutrition Board of the US National Academy of Sciences National Research Council recommended a change in the level of iron to meet the needs of the female population by increasing it by at least 5 mg/day, the form of iron being such as to ensure 10% absorption and that this could be done by increasing the levels of iron now prescribed by US federal standards for enriched cereal products: the desirable goal being 110 mg Fe/kg of enriched flour and 66 mg/kg of bread. Likewise, they recommended that other common farinaceous products be also included.

Table I. The ionic composition of sea water (Woods Hole) and the body fluids of various animals, relative to sodium as 100. Derived from VILLEE and DETHIER [1972]

	Ca	Mg	SO_4
Sea water	2.79	13.9	7.10
Aurelia, coelenterate	2.15	10.18	5.15
Stronglylocentrates, echinoderm	2.28	11.21	5.71
Phascolosoma, sipunculid	2.78	–	–
Venus, mollusc	2.17	5.70	5.84
Carcinus, crustacean	2.51	3.70	3.90
Hydrophilus, insect	0.92	16.8	0.12
Lophius, fish	1.01	1.61	–
Frog, amphibian	1.92	1.15	–
Man, mammal	1.78	0.66	1.73

A review of the report of the American Medical Association's Council on Foods and Nutrition [1972] on iron enrichment of wheat flour and baked goods indicates that two presently available iron compounds are preferential under specified conditions, namely reduced iron for flour and other products requiring extended shelf or storage life, and the smaller the particle the greater the availability, and ferrous sulphate (dried) which also has high bioavailability where the choice is for bread and other baked products where prolonged shelf life is not a factor. There is still need for one or more iron compounds where there is a more flexible range of usage compatible with bioavailability.

Man differs from the rat in respect of the amount of iron normally absorbed, namely about 3% as against 30% for the rat. In young Swedish soldiers, 77 mg Fe/day were consumed and this mainly in the form of haem. The absorption was close to 5.5% and presumably covered all requirements.

From studies on the absorption of iron by everted sacs, it would seem that iron passes by simple diffusion across the mucosal membrane into the epithelial cells from the intestinal lumen, there being a two-way movement of iron across the luminal membrane. Iron enters a labile pool in the epithelial cells while some passes into the plasma. Transfer to the plasma may be by diffusion in the iron-replete animal and may be bi-directional; in iron deficiency it may be increased by active transport. Iron not transferred to the plasma is stored temporarily in the epithelial cells and lost to the lumen by cell turnover. *In vivo,* it would be legitimate to believe that intestinal control over iron absorption is a delicate interplay of mucosal and serosal transfer of iron, iron deposition in the mucosal cells, and the turnover of these cells [O'BRIEN, 1973].

It is believed that in the rat there is a factor in gastric juice, other than acid, which plays a significant role in the absorption of ferrous salts as absorption could be doubled in rats with achylia gastrica (induced by 1,750 rad X-ray) if given neutralized human or rat gastric juice [MURRAY and STEIN, 1971].

It has been suggested from observations on jejunal contents of healthy volunteers that dietary non-haem iron when released from a meal is maintained in solution by binding to a large molecular weight mucopolysaccharide. Iron is gradually transferred from this carrier to a low molecular weight ligand which may faciliate absorption [GLOVER and JACOBS, 1971].

MCCANCE and WIDDOWSON [1938], in their classical experiments on iron absorption, demonstrated that the primary control of iron balance is by regulation of iron absorption since processes for enhancing secretion in man are quite limited. Total body iron stores and the rate of erythropoesis are the most important controlling factors, as already pointed out.

The availability of iron from sources used in the enrichment of bread was estimated in anaemic rats given enriched bread containing 20 ppm iron. The amount of Hb regenerated in 30 days from two reduced irons and from ferric orthophosphate was about two thirds, and that from sodium iron pyrophosphate about one third of that from ferrous sulphate. The earlier that the availabilities were estimated during the regeneration phase, the more pronounced were the differences. At day 15, the two reduced irons, ferric orthophosphate and sodium iron pyrophosphate were only 38, 28 and 6% as available, respectively, as ferrous sulphate; with higher iron intake, 110 ppm, the same iron sources were almost equally effective. Under simulated gastric conditions, the solubilities of test iron, except ferrous sulphate, correlated well with the results of bioassays. Tested in a separate experiment, ferric ammonium citrate was found to be better assimilated by anaemic rats than ferrous sulphate on diets containing 20 ppm iron. The process of baking seemed to increase availability partially [RANHOTRA *et al.*, 1971].

ASHWORTH *et al.* [1973] studied the absorption of iron from ^{59}Fe-labelled maize and soya bean preparations, as measured by whole-body counting in 42 apparently healthy Jamaican infants and compared the absorption with the absorption of ferrous ascorbate. The mean absorption of iron from maize was 4.3% and from soya beans baked at 300 °C, 9.4%; compared with 28.5% for ferrous ascorbate. In a group of children given boiled soya beans, the mean absorption of iron was 2.8% and of ferrous ascorbate 16.7%. The absorption of food iron was not increased in children who were considered to be anaemic (Hb less than 100 g/litre) or iron-deficient (serum iron less than 500 g/litre and saturation of total iron-binding capacity less than 15%). The poor availability of iron in maize meal – the staple food of children in Jamaica – is probably an important cause of the high prevalence of iron deficiency anaemia.

A collaborative study with eight laboratories by PLA and FRITZ [1971] was used to assess the availability of iron to chickens and to rats. The animals were considered depleted when their haemoglobins were less than 5 and 6 g/100 ml blood and haematocrits were less than 24 and 29%, respectively.

Relative to $FeSO_4 7 H_2O$ taken as 100, values were for ferric phosphate, 12 ± 10.9; sodium iron pyrophosphate containing 14.5% iron, 13 ± 7.7; reduced iron, 46 ± 17.9; and ferrous carbonate cres containing 38% iron, 3 ± 4.5. The substances were ranked in the same order when the plasma irons of volunteers were estimated. The Hb repletion test was recommended to the US Association of Analytical Chemists for adoption as official first action.

The nature of the lipaemia in iron deficiency anaemia has been further described. After 5 weeks, lipaemia was evident in rats on diets low in iron; it was characterized by a rise in plasma triglycerides. Plasma cholesterol and phospholipids were low or relatively unchanged. After 1 week of iron supplementation the rats showed the expected haematological response which was accompanied by a significant reduction in serum triglycerides. Triglyceridaemia associated with anaemia was dependent on the kind and amount of fat in the diet as well as the protein content. Severe iron deficiency in chickens associated with anaemia, was dependent on the kind and amount of fat in the diet as well as the protein content. Severe iron deficiency in chickens was also associated with a rise in serum triglycerides and smaller rises of cholesterol and phospholipids. This lipaemia was associated with a significant rise in the total plasma proteins, primarily albumin, and a moderate accumulation of fat in the liver [AMINE and HEGSTED, 1971].

The effect on rats rendered anaemic through weekly bleeding for 4 weeks was to induce a significantly lower tensile strength and a higher incidence of non-union of fractures and slower ossification of the cartilaginous callus than in the controls. The adverse affect of this anaemia on wound healing was attributed mainly to decreased tissue oxygen although tissue deficiency in iron may have been a contributing factor [ROTHMAN *et al.*, 1971].

Chronic iron deficiency in rats may lead later to demineralisation of bones (osteoporosis), not obvious in controls given added ferrous ammonium sulphate [SCHMIDT *et al.*, 1971].

The renewed interest in the role of copper in iron metabolism has been identified at the molecular level as the copper protein ceruloplasmin. It has been shown by FRIEDEN [1970] that both *in vivo* and *in vitro* the rate of formation of Fe (111) transferrin, and ultimately of haemoglobin biosynthesis is conditioned by this molecular link which appears to be directly related to the ferroxidase activity of ceruloplasmin. The parallel observations of LEE and MATRONE [1969] have shown that there is likewise a relation between ceruloplasmin and haemoglobin synthesis in which excess zinc somehow interferes with utilization of copper in the biosynthesis of ceruloplasmin.

VIII. Factors Influencing Requirements of Domestic Birds

A. Calcium and Phosphorus

The requirements for both calcium and phosphorus are influenced by the level of dietary vitamin D and, in general, the requirements for both of these elements increases as the level of this vitamin decreases and *vice versa.* It is obvious that there are levels of these minerals below which no amount of vitamin D will compensate (table II).

It is suggested that it is the gut contents of invertebrates, such as caterpillars, insects and earthworms, rather than their tissues which provide nestling wild birds, like blackbirds and thrushes, with their calcium, which increases some 100 times over the amount at hatching [BILBY and WIDDOWSON, 1971].

The UK Agricultural Research Council survey [1975] of the literature has indicated that the availability of phytate-P for young chicks and laying hens (but not for growing pullets) is negligible at dietary levels of calcium high enough to permit maximum bone or egg shell calcification, although it can be considerable at lower levels of calcium. The requirements for Ca, P, Mg and Fe are set out in table II, and these recommendations are based on the proposition that phytate-P is considered completely unavailable when practical diets are being formulated for chicks or laying hens. As calcium phytate is very insoluble and calcium bound in this form is poorly available, so the requirements for calcium increases.

Table II. Domestic birds. Summary of requirements for Ca, P, Mg and Fe, expressed as g/kg of diet containing 3.0 Mcal (12.6 MJ) ME/kg, unless otherwise stated. Based on UK Agricultural Research Council [No. 1, 1975] estimates. Values in italics represent preferred estimates

	Ca	Non-phytate-P	Total P	Mg	Fe
Fowls					
0–4 weeks	11.5–12.0[1]	4.6–4.8		0.4	
4–8 weeks	*7.0–8.0*		*5.7–6.3*	*0.4*	*75 mg*
8–18 weeks	*4.0*		3	0.4	
Layers and breeders					
On wire	3.8–4.0	0.35–0.40 g/day		*0.4*	
On litter	g/day	0.30–0.35 g/day			
Turkeys					
0–8 weeks	7.5–8.5[1]		6–7		
8–20 weeks	4.0–6.0		4–6	*0.45*	
Breeders	5.0–6.0 g/day		5		
Ducklings					
0–4 weeks	5.6			0.5	
Goslings					
0–6 weeks	4.0		4.6		
Pheasants					
0–5 weeks	10		8		
5–14 weeks	5		4		
Japanese quail					
0–6 weeks	5		4		

1 Provided diet does not contain more than 2 g phytate-P/kg.

1. Growing Poultry

It is generally found that greater amounts of dietary calcium are required for maximum bone ash percentage than for maximum growth, although the requirements for phosphorus is similar for both.

To obviate the suggestion of precision implicit in a single figure, but at the same time to discourage the use of excessive amounts of minerals in the diet, a range of values is provided in most cases.

In assessing the requirements for older growing pullets, we take into account their subsequent laying performance.

The greatest single factor responsible for variation in previous assessments has been the metabolizable energy content of the diets. For this reason the requirements have been estimated on this basis. Where the experimental data have permitted the construction of dose response curves, these have been used by the UKARC in estimating requirements, but it must be admitted that these curves are very frequently of the 'diminishing response' type to further increment and showing no genuine plateaux.

2. Laying Hens

Egg production and shell thickness, in general, are the criteria used. On the whole, the requirements for the laying bird for most nutrients are seldom more than two or three times that of non-laying birds of the same age, but for calcium the requirements are 20–30 times greater for the laying over the non-laying bird. As voluntary intake is related to live weight, laying diets set out for small birds must contain higher levels of calcium than those designed for heavy birds. It is for this reason that the assessed requirement is set forth per bird per day during egg shell development. Laying hens consume approximately 20% more feed on egg-forming days [NORRIS and TAYLOR, 1967].

Compared to calcium, the requirements of phosphorus for egg production are fairly small. When lower concentrations are included in the feed and the balance is provided by a form of calcareous grit, it is not possible to ration the birds with accuracy and a generous excess of grit must be given. The system of giving all the calcium in the diet is becoming increasingly common, but has not yet been fully evaluated against the traditional method of supplying calcium. If calcareous grit is being provided it would not appear to be any advantage in including in the diet more calcium than is required on days when egg formation is not taking place, possibly 10 g Ca/kg diet.

3. Turkeys

Turkey poults are extremely sensitive to a deficiency of phosphorus, and an excess of calcium can readily depress growth by inducing a phosphorus deficiency. The phosphorus requirement is further complicated by the fact that certain sources of supplementary phosphorus are much less available than others. Thus, potassium dihydrogen orthophosphate, primary calcium orthophosphate and hydrated secondary calcium orthophosphate are highly available, while tricalcium phosphate and anhydrous secondary calcium phosphate show much lower availability [GILLIS *et al.*, 1962; SCOTT *et al.*, 1962]. Steamed bone flour is presumed to be in this category. A further complication is that turkey poults grow less well on purified diets than on diets containing a proportion of natural feed ingredients, such as soya bean meal or soya protein. Other factors in soya beans increase the availability.

B. Magnesium

1. Growing Poultry

Conventional feeds contain adequate amounts of magnesium and no supplementation is necessary or indeed desirable. When purified diets are used, a requirement of between 0.35 and 0.40 g/kg is suggested. As 5 mg/kg are required for each additional 1 g/kg phytate phosphorus, such a correction is needed.

For young turkeys it is suggested that 0.45 g/kg be taken as a provisional requirement being a compromise between the findings of KEENE and COMBS [1962] and SULLIVAN [1962].

The magnesium requirements of older growing chickens and turkeys has apparently not yet been determined. The magnesium requirements would appear to be around 0.48 g/kg [VAN REEN and PEARSON, 1953].

For laying and breeding hens a provisional requirement of 0.4 g/kg has been recommended. There is no information on breeding turkeys.

C. Iron

With an adequate level of iron in the diet, absorption is in the region of 40–50% in very young chicks but only 5–10% in older birds.

It has been frequently shown that calcium as calcium carbonate and also phosphate as the ammonium salt, given separately, increase the requirement for iron as judged by their effects on haemoglobin concentration [WADDEL and SELL, 1964]. There is no evidence that soya bean protein reduces the availability of iron as it does with zinc.

The UKARC [1975] have suggested 75 mg/kg feed which is based on the data of DAVIS *et al.* [1968] and the requirement of the poult is probably similar. No data are available for adult birds.

IX. Factors Influencing Requirements of the Major Elements by Mammals – Calcium, Phosphorus, Magnesium, Iron and Sulphur

In general, the requirements of minerals are assessed by determining the accretion of the element in growth, and compensatory effects for metamorphosis in species where this happens, then for maintenance in the adult, but in the case of the female there is also the accretion during the reproduction phase and in mammals also the excretion of the element in milk. Endogenous losses are computed and the availability of the element is assessed.

In the case of farm animals, the most recent practice has been a twofold: firstly, by a factorial method the minimal requirements of animals of different classes, producing at different rates, are calculated, and secondly, these factorial estimates are compared with the results of experiments and feeding trials in which the element has been given in two or more different amounts and the resultant performance measured. The factorial method assesses requirements of animals in two stages. In the first, the net requirement of the animal is obtained from estimates of the storage and excretion of the element made during growth, pregnancy and lactation, and any inevitable losses of the element from the body (the endogenous losses). Secondly, the availability of dietary sources of the element is assessed from metabolism experiments. The net requirement divided by the availability gives the requirement in terms of a dietary amount [UKARC, 1965].

For net requirements of each element see list on opposite page.

The accuracy of the factorial method is limited by the amount and accuracy of the basic information concerning the individual items. In the case of the minerals, we are concerned with the accuracy which is almost entirely determined by the low accuracy of the availability of mineral elements in rations. Thus, if the availability of a mineral in a food item appears to be 20–25%, the dietary needs are about four times the requirement if the true availability is 25% but five times the requirement if it is 20%. The main discrepancy is frequently to be found in the assessment of endogenous losses rather than in the measurements of accretions and secretions. This is largely due to paucity of data on endogenous losses. There is also a lack of precision in net requirements for growth as there is insufficient data on the mineral content of animals growing at different rates at different ages.

If practical trials indicate that factorial estimates of requirement are incorrect, in particular too low, then it is essential to re-examine the bases of these requirements and to accept as requirements values based on practical tests.

Net requirement for growth (G)	= daily retention during each particular phase of growth
Net requirement for pregnancy (P)	= daily retention of specified stage of pregnancy
Net requirement for lactation (L)	= daily excretion in the milk yielded
Net requirement for maintenance (M)	= daily excretion during a given stage of body weight maintenance with normal performance; during growth the maintenance requirement cannot properly be measured as they are metabolically indistinguishable; for some purposes the excretion during arrestment of growth will provide a figure for maintenance of that phase
Net endogenous requirement (E)	= the inevitable loss of the element in faeces and tissue which occurs even when the element is absent from the diet; there is sometimes included epidermal losses as in sweat and hair and cutaneous tissue
Total net requirement for the reproductive female	= G + P + L + E
For the adult organism in weight equilibrium and normal performance	= M + E

A. Major Minerals in Milk of Different Species

Despite a large variation in the quantity secreted for unit volume, the composition of the ash fraction of milks of different species shows relatively little variation (table III).

BUNGE [1874] in his clinical studies of milk composition showed that the composition of the ash of milk was very similar to the composition of the ash of the embryo, though the ash of embryos and new-born animals usually contain considerably more phosphorus than does milk ash.

Species less mature at birth, such as the rat, the rabbit and the dog, produce milk ash containing more bone forming elements, but differences are small, and we must note that the sow produces milk with an ash containing 26% calcium for its piglets which are born well calcified, whereas the rabbit, which is poorly calcified at birth, produces milk of about the same percentage of calcium in its ash.

Table III. Mineral constituents of milk of different species of mammals. Compiled from various sources, principally LING *et al.* [1961]

Species	Time in days to double bulk weight	Total ash, g/100 ml	Ca, g/100 ml	P, g/100 ml	Mg, g/100 ml	Fe, µg/100 ml
Man	180	0.2	0.03	0.014	0.004	150
Ass	–	–	0.09	0.05	–	–
Horse	60	0.4	0.10	0.06	0.01	–
Ox	47	0.7	0.12	0.10	0.12	100
Goat	19	0.8	0.14	0.12	0.15	–
Pig	18	–	0.27	0.16	0.09	180
Buffalo	–	–	0.18	0.12	0.06	–
Sheep	10	0.9	0.19	0.15	0.14	–
Camel	–	0.9	–	–	–	–
Reindeer	–	1.4	–	–	–	–
Dog	8	1.3	–	–	–	900
Cat	7	–	–	–	–	–
Rabbit	6	2.2	0.55	–	–	120
Rat	6	2.0	–	–	–	700

B. Calcium

1. Accretion of Calcium during Growth

The calcium contents of individual tissues of the body vary considerably. In higher animals, the muscle mass contains about 100 mg Ca/kg fresh weight, the fat depots virtually none while the bones contain 110–200 g/kg. Unless the soft tissues, the bones and the fat depots increase in weight at the same relative rates throughout growth, the calcium content of a gain in body weight cannot be constant. As an animal ages, its gains in weight for a considerable time contain an increasing proportion of fat. Furthermore, at a given age, large weight gains contain proportionally more fat than do small ones. Again, although skeletal growth rate falls markedly with age, there is ample evidence that in young animals the skeleton becomes increasingly mineralized with increasing age, until at maturity increases in body weight are not necessarily associated with changes in calcium deposition in the skeleton.

It is obvious that the calcium content of a gain in body weight can vary considerably and, unfortunately, there are few data even on farm livestock on which to make accurate predictions. For ruminants with no allowance for breed differences, the following quadratic equation relating total body calcium (g) weight (W, kg) has been evolved from data in the literature [UK Agricultural Research Council, 1965]: $Ca = -483 + 16.68\ W - 0.0072\ W^2$.

2. Deposition of Calcium in Growing Animals

The calcium content of the body of mature sheep (8.9 g/kg) is lower than that of mature cattle (12 g/kg body weight). This real difference is not as one might have expected due to an association of body composition with body size since the mature rat contains 11–14 g Ca/kg body weight a content much the same as that of adult cattle.

3. Calcium Deposition in Pregnancy

Calcium is deposited in the fetus during pregnancy and in lesser amounts in the placenta. In addition, the uterus of the mother increases in size, there is growth of her mammary glands and usually she increases in weight. At birth blood is lost. Thus, the total requirement for calcium during pregnancy consists of the calcium content of the fetus at term, the calcium content of the adnexa and the maternal storage of the element in new growth.

Young animals continue to grow during pregnancy and the net requirements for pregnancy are in addition to those for any normal growth made. Even older animals show considerable maternal gains. MORGAN and DAVIS [1936] have measured the total gain, maternal gain and gain in weight of the products of conception of Ayrshire cows during 1st, 2nd, 3rd, 4th to 11th pregnancy. The estimate of the amount of calcium in the non-fetal reproductive structures has been based on the data of FORBES *et al.* [1935] for the calcium content of the placenta and fetal fluids and an assumed value of 0.15 g/kg for the calcium content of the wall of the uterus.

a) Deposition of Calcium during Pregnancy in Sheep

On average, 89 g calcium are lost to the body at lambing: this applies to ewes most of which had twins [UK Agricultural Research Council, 1965]. If it is assumed that the calcium content of the fetus is the same as that of the lamb at birth, namely 10.5 g/kg, and that the calcium content of the uterus and placenta is 150 mg/kg and that of the fetal fluids 50 mg/kg, corresponding to a plasma ultrafiltrate, then an estimate of the calcium deposition in g/day rises to 1.43 g in singles and 2.18 g in the case of twins. As young sheep grow during their first pregnancy it is assumed that the gains have a calcium content of 8.9 g/kg [UK Agricultural Research Council, 1965].

4. Calcium Excretion in Lactation

The calcium excreted in milk is the product of milk yielded and the calcium concentration of that milk. Yields vary widely according to breed and from circumstance to circumstance. Calcium is more closely related to the protein content of milk than to the fat but the latter is more usually available. For the cow: calcium content (%) = 0.94 + 0.085 (fat content, %).

The average content of calcium in ewe's milk is appreciably higher than that of cow's milk. The output of calcium in the milk (g/day) of ewes, and assuming a calcium content of 1.9 g/kg milk, is up to 1.5 g/day.

The main routes of excretion of calcium, magnesium and iron, are in the faeces and those of phosphorus and sulphur in the urine. This also relates to the pig.

X. Requirements of Ruminants for Calcium, Phosphorus and Magnesium (tables IV–IX)

The UK Agricultural Research Council [1965] pointed out that the amount of calcium and to a lesser extent the amounts of phosphorus and magnesium deposited in each kg of weight gained by ruminants during growth depend on the age, growth potential and previous nutritional history of the animal. This is due to the fact that the skeleton and soft tissues can and do grow at different rates. Although this applies strictly to cattle it is presumed that comparable indications for sheep, now presently lacking will substantiate this generalization. In addition, there is considerable evidence that the availability of these three elements in the diet of cattle decline as the animals matures.

A. Pregnancy

Dietary requirements for pregnancy in cattle have been based on determination of the amounts of mineral elements retained in the fetus and adnexa, these being added to the amounts required for maintenance. With pregnant heifers, a small additional allowance has been included for their own growth. The requirements for mature cattle can be applied to heifers with little error, simply by assuming that the requirements of a heifer are those of a mature animal weighing 100 kg more than the heifer.

Table IV. Dietary calcium, phosphorus and magnesium requirements of growing cattle, when the rates are different [UKARC, 1965]

Body-weight, kg	Rate of gain (g/day) at 0.33 kg/day			Rate of gain (g/day) at 0.5 kg/day			Rate of gain (g/day) at 1.0 kg/day		
	Ca	P	Mg	Ca	P	Mg	Ca	P	Mg
50	6.7	4.3	0.4	9.6	6.2	0.5	18	12	0.8
100	11	5.4	1.1	15	7.3	1.2	27	13	1.7
200	14	8.1	3.7	18	9.8	4.0	30	15	5.0
300	18	13	5.2	21	15	5.5	33	20	6.5
400	22	22	6.7	26	24	7.0	37	29	8.0
500	25	28	8.2	28	29	8.5	38	33	9.5

Table V. Dietary calcium, phosphorus and magnesium requirements of cattle during pregnancy (mature cow weighing 500 kg) [UKARC, 1965]

Stage of pregnancy, months	Ca, g/day	P, g/day	Mg[1], g/day
5–6	20	26	7.9
6–7	27	30	8.0
7–8	35	34	8.8
8–9	33	34	9.4
Non-pregnant	18	26	7.5

1 It is recommended that this minimal allowance should be increased by 2.0 g to take into account variation in need from individual to individual.

Table VI. Dietary calcium, phosphorus and magnesium requirements of growing sheep (30–50 kg body wt) when the rates are different [UKARC, 1965]

Rate of growth, g/day	Requirement, mg/kg body wt/day		
	Ca	P	Mg
50	110	65	14
100	130	75	16
200	180	90	21

Table VII. Dietary calcium, phosphorus and magnesium requirements of pregnant ewes (50-kg ewe) [UKARC, 1965]

Month of pregnancy	Ewes bearing single lambs			Ewes bearing twins		
	Ca, g/day	P, g/day	Mg, g/day	Ca, g/day	P, g/day	Mg, g/day
2	4.5	3.7	0.59	4.5	3.7	0.59
3	4.8	3.8	0.61	5.2	4.0	0.65
4	5.8	4.2	0.68	7.1	4.8	0.80
5	7.6	5.1	0.85	9.3	5.8	0.98

Table VIII A. Calculated dietary requirements of milking cows for calcium, phosphorus and magnesium (g/day), based on the dietary requirements given in table VIII B [UKARC, 1965]

Milk yield, kg	Friesian cow, weight 500 kg, 3.8 % fat in milk			Jersey cow, weight 350 kg, 6.0 % fat in milk		
	Ca	P	Mg	Ca	P	Mg
5	32	34	11	29	23	8
10	46	43	14	44	32	12
15	60	51	17	60	40	15
20	74	60	20	76	49	18
25	88	68	23	92	57	21
30	102	77	26	108	66	24

Table VIII B. Dietary requirements

	Calcium	Phosphorus	Magnesium
Friesian maintenance, g/day	17.8	25.5	7.5
Jersey maintenance, g/day	12.4	14.9	5.3
Friesian milk, g/kg/day	2.8	1.7	0.63
Jersey milk, g/kg/day	3.2	1.7	0.63
Availability, %	45	55	20

It is recommended that these minimal allowances for magnesium should each be increased by 2.0 g to take into account variation in need from individual to individual.

Table IX. Dietary calcium, phosphorus and magnesium requirements of lactating ewes (maintenance and lactation) [UKARC, 1965]

Month of lactation	Lowland breeds, 50-kg ewe			Hill breeds, 35-kg ewe		
	Ca, g/day	P, g/day	Mg, g/day	Ca, g/day	P, g/day	Mg, g/day
1	14	9.3	2.6	11	7.0	2.0
2	12	8.3	2.2	9.3	6.2	1.8
3	10	7.0	1.7	7.6	5.2	1.4
4	7.8	5.7	1.3	5.8	4.2	1.0

The requirements of pregnant sheep obviously vary with the number of lambs the ewe carries. The difference in the dietary needs of ewes bearing single lambs and those bearing twins, however, is not large and is apparent only in the final stages of pregnancy.

B. Lactation

The requirements of the lactating cow for calcium appear to vary with both the age of the cow and the fat content of her milk. Thus, the calcium content of milk increases by 20% when the fat content increases from 3 to 5%, but at the same time the phosphorus content increases by 8%. Age rather than mature size of animal appears important in that there is evidence, not restricted to that from cows, that the availability of dietary calcium declines with advancing age. The calculated requirements for calcium, phosphorus and magnesium of Friesian cows (500 kg body weight) and of Jersey cows (350 kg body weight) for yields ranging from 5 to 30 kg milk/day are given in table VIII A. It is recommended that the allowances for magnesium should be increased by 2.0 g/day, because the requirement we have calculated is an average estimate of minimal needs, and individuals can well have slightly higher requirements.

With lactating sheep estimates have been made of the milk production likely under two different sets of circumstances. The results given in table V apply to lowland breeds producing 2.5 kg/day at the peak of lactation and to hill breeds producing 1.8 kg milk/day at that time.

C. Antlers

In a personal communication from Dr. R. N. B. KAY of the Rowett Research Institute, Aberdeen, it would seem that from their analyses a prime stag growing 4 kg of antlers (dry matter) in 3 months (May–July) in North East Scotland requires a total of 600 g Ca, 300 g P and 12 g Mg. This requirement for calcium would appear to be almost twice that required by a hind during a full lactation (160 kg in 150 days) when 350 g Ca, 350 g P and 30 g Mg may be produced in the milk [ARMAN *et al.*, 1974].

D. Sulphur in Ruminant Nutrition

WHANGER [1972] has recently reviewed this subject and points out that with the increasing use of non-protein nitrogen (NPN) as a major source of supplementary nitrogen in ruminant rations, there is a greater possibility of sulphur deficiency particularly where poor quality roughages, or roughages grown on sulphur-deficient soil, or these roughages fed with NPN are the main source of ruminant feed.

Although mammals can oxidize reduced sulphur compounds and incorporate sulphate into various molecules, they are unable to reduce to the sulphide level [HUOVINEN and GUSTAFSSON, 1967] and must depend on plants and bacteria to provide them with reduced sulphur compounds. Because the rumen microorganisms can reduce oxidized forms of sulphur to forms which can be incorporated into organic forms, ruminants have the ability to obtain their sulphur form inorganic sources of sulphur [ANDERSON, 1956]. Hydrogen gas appears to serve as the best hydrogen donor in sulphate reduction. Sulphur amino acids are synthesized by those microorganisms in the rumen and incorporated into their own proteins which are later digested in the further stages of gastro-intestinal digestion. It has been estimated that about 40% of infused ^{35}S in the form of $Na_2\,^{35}SO_4$ was cycled to the rumen and incorporated into microbial protein.

It had been suggested that molybdate and sulphate might be responsible for restricting copper utilization by depressing copper solubility in the digestive tract through formation of cupric sulphide, but recent observations by MILLS [1975] suggest that most experimental conditions have involved too high molybdate concentrations. Molybdate appears to inhibit reduction of sulphate by rumen microorganisms and copper decreases this inhibition by molybdate [PRENTICE and MATRONE, 1970]. Recent research has confirmed that copper plays an intrinsic role in utilization of sulphur for wool fibre formation [KAPOOR *et al.*, 1972].

XI. Factors Affecting Requirements of Horses for Major Minerals

From a review of the literature, OLSSON [1969] concluded that for full-grown horses on maintenance and for working horses, the respective intakes of calcium and phosphorus per day should be around 25 and 17 g for maintenance, and 35 and 25 g for working. The following respective figures for calcium and phosphorus also apply: for late pregnancy, 60 and 33 g; for lactation, 68 and 45 g; for growing foals at 6 months, 35 and 21 g, and at 12 months, 39 and 25 g.

OLSSON also subscribed to the US National Academy of Sciences' publication No. 912 [1961] view that any ration containing at least 50% of forage may be expected to meet the magnesium and potassium needs of horses.

There is no evidence that in horses, inorganic sulphur in the feed is used for the synthesis of sulphur-containing amino acids. It is considered that a sulphur intake corresponding to 0.15% of the dry matter is adequate.

As for iron, it is generally supposed that horses ordinarily fed with green fodder, hay and grain, in general are supplied with 80 ppm or more iron in the dry matter, which is in excess of requirements.

MÜLLER-REH [1972] has conducted a very extensive study of the intakes of minerals by altogether 184 breeding, riding or jumping horses of different ages in 18 groups from different establishments in the German Federal Republic. They were studied in winter when stabled and in summer when they were on pasture. In winter, when the food was oats and hay, sometimes with pelleted or loose concentrate and mixture of minerals and vitamins, the daily intake of calcium in the individual stables ranged from 19 to 35 g and were often below 30 g. Those of phosphorus were always sufficient, ranging from 28 to 29 g. The Ca:P ratio was usually about 0.7; magnesium was from 10 to 19 g and always met requirements. Iron ranged from 11 to 31 mg/day and was considered sufficient. In summer, the percentages of the minerals in the pasture, but not the absolute amounts, were recorded. The calcium, phosphorus and Ca:P ratios were adequate and balanced, but magnesium did not reach normal requirements, and on some pastures iron was excessive. There was no correlation between intake and blood levels except in the case of iron. HINTZ and SCHRYVER [1972a] estimated the calcium and phosphorus requirements of ponies to be, respectively, 27.5 and 11.5 mg/kg absorbed.

HINTZ and SCHRYVER [1972b] also assessed the magnesium maintenance requirement of ponies to be 12.8 mg/kg. HINTZ *et al.* [1973] concluded from their examination of the availability of phosphorus in wheat bran that its phosphorus content was about half as available as phosphorus in supplements such as NaH_2PO_4, dicalcium phosphate or bone meal, and differences in availability should be considered when balancing rations for horses.

XII. Requirements of Laboratory Animals

1. Rats. One of the most recent assessments of the requirements of rats for the major minerals in that of SHTENBERG and TORCHINSKII [1972] who, having reviewed the literature, come to the conclusion that for an artificial diet based on casein, sugar, vegetable oil and vitamins, and in the following elements with which this paper is concerned, should be supplied in appropriate form in the following concentrations: S, 0.100%; Ca, 0.907%; P, 0.606%; Mg, 0.61%, and Fe, 0.005% of the total weight of the diet. Details are given for the other mineral elements. BELL *et al.* [1941] found 0.36% calcium quite adequate for strength of femora.

The USNRC Publication 990 specified the requirements by weight rather than as concentrations in the diet. Their values were: for growth – 0.060 g Ca, 0.050 g P and 4 mg Mg; for pregnancy – 0.12 g Ca, 0.10 g P and 10 mg Mg; and during lactation – 0.18 g Ca, 0.15 g P and 15 mg Mg.

2. Mice. The above reference also gives levels for calcium and phosphorus during growth as 0.021 and 0.018 g/day, respectively.

3. Guinea pigs. For calcium, phosphorus and magnesium, the recommendations on the foregoing publication are given as 0.1 g, 0.05 g and 28 mg/100 g body weight, respectively.

4. Hamster. The levels provided are 0.6 g Ca and 0.35 g P/100 g diet.

5. Dog. The USNRC data in Publication 989 [1962] are as follows:

	Body weight, kg	Ca, g	P, g	Mg, mg
For growth	2.3	1.2	1.0	50
	4.5	2.4	2.0	100
For adult	6.8	3.6	3.0	150
maintenance	13.6	7.2	6.0	300
	> 22.7	12.0	10.0	500

XIII. Requirements of Pigs

These have been set out in the UK Agricultural Research Council [1967] assessments (table X). They have been expressed as concentrations of the dry matter and in that publication there are given the approximate dry matter intakes per day.

The recommendations for calcium and magnesium for young pigs apply to diets with an availability of about 80%, 70% in the case of phosphorus. Those for growing pigs assume a value between 70 and 45% as they grow older. There is evidence that the percentage of calcium absorbed from any particular diet falls progressively with increase in age.

Table X. Requirements of pigs for Ca, P, Mg and Fe stated as g/kg dietary dry matter [UKARC, 1967]

Mineral	Live weight (kg) or class of pig	Estimated requirement, g/kg dietary DM	Toxic level
Ca	up to 20	8	1.2% when the zinc content of the diet is less than 50 mg/kg dry matter
	20–55	8	
	55–90	6	
	sows	6	
P	up to 20	7	
	20–55	6	
	55–90	5	
	sows	5	
Mg	up to 55	0.4	
Fe	up to 20	0.06	5 g/kg
S	it is the intake of the S-amino acids, methionine in particular which is the determinant of adequacy		

These figures are based on availabilities of Ca and Mg of around 80% and for P of 70%. The availabilities tend to decrease as animals get older.
The level of Fe recommended is intended to ensure haemoglobin levels of approximately 8 g/100 ml blood.
Excess of Ca and Mg can reduce the amounts of Fe and Zn absorbed.

The amount of calcium required by the pig also depends upon the presence in the feed of other dietary factors, notably phosphorus and vitamin D, so adequate amounts of these nutrients must be offered. The daily requirement to ensure maximum growth rate is considerably below that needed to ensure good bone formation.

Comparatively little is known about the requirements of the sow for calcium, though satisfactory results are obtained by feeding diets of the type suitable for pigs between 45 and 90 kg live weight. These diets lead to good bone formation in the progeny and ensure satisfactory lactation without imposing too great a strain on the sow's own mineral reserves.

Phosphorus requirements are closely associated with those of calcium. Indications are that phytate-P is somewhat less available than most inorganic sources of phosphorus. An excess of phytic acid should be avoided as this can reduce zinc absorption. The amount of phosphorus required for good bone quality, reproductive performance and lactation are higher than those needed to obtain maximum growth rate.

It has been found that about 400 mg Mg/kg dry matter in the diet is required for young growing pigs to ensure adequacy when the availability of the element is about 80% [ULLREY *et al.*, 1962].

The level of iron recommended (about 60 mg Fe/kg dry matter) should ensure for young pigs haemoglobin levels of approximately 8 g/100 ml blood. Excess of calcium, magnesium and manganese can reduce the amounts of iron and zinc absorbed.

XIV. Requirements of Subhuman Primates for Major Minerals

With the exception of the leaf-eating species, most of the Old and New World monkeys are omnivorous. Even the howler, woolly and spider monkeys occasionally eat small animals. Baboons are primarily vegetarians. The predominant diet of the chimpanzees is fruit. The diet of most gorillas usually consists of bamboo shoots, wild cherry leaves and fruit.

A. Calcium and Phosphorus

The data on calcium absorption, turnover and excretion were found to be similar to those in children, in contrast to studies on rodents [KERR, 1972]. Anticipating some of the problems that confront man during prolonged periods of space travel, KAZARIAN and VON GIERKE [1969] and PYKE *et al.* [1968] studied the effect of restraint on bone density and calcium and phosphorus metabolism of subhuman primates. Loss of bone strength, thinning of trabeculae, and resorption of cortical bone at the sites of muscle and tendon attachment were reported. Even with a diet high in calcium and phosphorus and an apparent decrease in bone density, the problem of elevated levels of faecal calcium and faecal and urinary phosphorus were still evident.

B. Magnesium

VITALE *et al.* [1963] produced magnesium deficiency by use of a purified diet in growing cebus monkeys. He got loss of weight, irritability, convulsions, reduced serum potassium levels and increased susceptibility to the toxic action of a cardiac glycoside. A marked increase in aortic lipid deposition was noted. They concluded that 40 mg Mg/kg/day was adequate to prevent the disorder.

C. Iron

FITCH *et al.* [1964] have concluded that soy protein, in some way, reduces the intestinal absorption of iron in the monkey.

KERR [1972] has listed four commercial chow diets – their contents range: Ca, 0.86–1.02%; P, 0.47–0.76%; Mg, 0.11–0.16%; Fe (ppm), 76–300; methionine, 0.39–0.52%, and cystine, 0.78–0.41%.

XV. Requirements of Man

The 8th edition of the US National Academy of Sciences [1974] 'recommended dietary allowances for man' has recently been published and there the factors influencing nutritional needs are discussed together with the precautions that should be exercised in using the 'allowances'. For the major minerals here considered, the level of nutrient intake judged to be adequate is that which meets the physiological needs of most healthy persons – a public health concept. The 'allowances', which are average values, must therefore exceed average nutrient requirements. Thus, a person consuming a diet containing less than that recommended is not necessarily consuming an inadequate amount of that nutrient. Nevertheless, the greater the proportion of people in a population group who habitually have intakes of nutrients below the 'recommended dietary allowances', the greater the risk that some of them will be consuming diets that are nutritionally inadequate.

These *dietary* allowances are not to be confused with the 'US recommended *daily* allowances' (RDA) proposed by the Food and Drug Administration [1973] as standard for nutritional labeling of foods. The RDA are based on the 1968 recommended dietary allowances but are, of necessity, limited to a few broad age categories. The values selected usually represent the highest allowance for any age group within the broader age category.

Table XI sets out the US National Research Council 'recommended dietary allowances' [1974], and where appropirate the UK Department of Health and Social Security [1969] 'recommended intakes of nutrients for the United Kingdom' are set alongside, together with the actual intakes revealed by the UK National Food Survey.

It has been customary to define a man's requirement for calcium and phosphorus through balance studies. But such techniques are an approximation and with one other exception cannot be recommended as a means of finding the other mineral element requirements of man. The technique is very difficult through problems of sampling, and until metabolic pathways of utilization and excretion are fully understood, interpretation will be questionable. However, substantial progress has been made in the study of iron requirements using balance studies, but in this area the main advance has been in measuring the availability from food labelled with radioactive iron.

Information on the calcium, phosphorus, magnesium and iron contents of components of human diets is readily available but the concentration of many of these elements in many drinking waters from different sources is practically unknown. It has been suggested that osteoporosis may be due to impaired vitamin D metabolism directly affecting bone.

One of the most controversial problems is the evaluation of iron and the recommended intakes. There is general agreement about losses of iron from the gastro-intestinal and urinary tracts and from skin (amounting to 14 μg/kg/day) and for normal women during menstruation (amounting to about 2 mg/day); the net loss of iron during pregnancy amounts to about 570 mg, which includes loss of blood at delivery and in the puerperium, as well as loss to the fetus and placenta, in addition to basal losses (via the skin, etc.) [ARC/MRC, 1974]. There is some disagreement amongst experts as to the

Table XI. US National Research Council recommended daily dietary allowances for Ca, P, Mg and Fe – revised 1974 (designed for the maintenance of good nutrition of practically all healthy people in the USA), and alongside the UK Department of Health and Social Security [1969] recommended daily intakes

Age (years) or condition	Ca, mg		P, mg		Mg, mg		Fe, mg	
	USA	UK	USA	UK	USA	UK	USA	UK
Infants								
0.0–0.5	360	600[1]	240	300–	60		10	6
0.5–1.0	540		400	470	70		15	
Children								
1–3	800		800		150		15	7
4–6	800	500	800	500	200		10	8
7–10	800		800		250		10	10
Males								
11–14	1,200	700	1,200	700	350		18	13–15
15–18	1,200	600	1,200	600	400		18	
19–22	800		800		350		10	
23–50	800	500	800	500	350		10	10
51 +	800		800		300		10	
Females								
11–14	1,200	700	1,200	700–	300		18	
15–18	1,200	600	1,200	600	300		18	12
19–22	800		800		300		18	
23–50	800	500	800	500	300		18	
51 +	800		800		300		10	10
Pregnancy	1,200	1,200	1,200	1,200	450		18[2]	15
Lactating	1,200	1,200	1,200	1,200	450		16	15
UK ordinary diet provides for adult		1,000		1,200		150–450		12.5

S related essentially to requirements for S-amino acids.

1 Artificially fed.

2 This increased requirement cannot be met by ordinary diets; therefore, the use of supplemental iron is recommended.

amount of iron needed by infants, children and adolescents not only to replace that loss, but also to provide for the increase in total haemoglobin, and in the iron content of tissues and to build up iron reserves. The absorptive capacity, as we have seen, is dependent on the individual's iron status.

According to the ARC/MRC Committee [1974], the UK's iron fortification policy with respect to bread flour is, to a large extent, ineffective. In view of the finding that ascorbic acid has some effect in promoting iron absorption, it has been claimed that in the treatment of iron deficiency anaemia, increased intakes of vitamin C may be more valuable than iron supplements.

It has been suggested that under the sedentary conditions of modern life, a moderate reduction of haemoglobin below the present accepted standards of normal causes no fundamental impairment and may even have advantages in terms of longer survival and low morbidity from cardiovascular disease [ELWOOD, 1974], but much more observation is required.

Summary

Little is known concerning the requirements of marine animals for the major mineral elements calcium, phosphorus, magnesium, sulphate and also for iron, although much is known about their osmotic regulation, rates of accretion and periodicity. Groups of invertebrates in which calcium is obviously very necessary are the molluscs for their shells, crustacea for their carapace and corals for their skeleton. In marine animals, the main source of these major minerals is by direct absorption from the environment: their actual ingestion is not necessary, for exchange of ions occurs across body walls, skin and gills. However, phosphorus and sulphur are obtained probably more effectively by fishes from their natural food. Where this is in poor supply, and particularly in soft waters, it may be necessary to fertilize the water with the needed minerals or suitably supplement the mineral content of the food fed to the fish to a concentration somewhat comparable to that used for small laboratory animals. In experimental diets for insects, it is usual to provide for minerals by inclusion of a popular mammalian salt mixture, but some of the mixture may only be required in traces, e.g. calcium.

The metabolism of these mineral elements in relation to dietary requirements as observed in birds and mammals – particulary those of agricultural importance and also of laboratory animals – is discussed, and also those of subhuman primates and of man himself by age group.

References

AGRELL, I.: The aerobic and anaerobic utilization of metabolic energy during insect metamorphosis. Acta physiol. scand. *28:* 306–335 (1953).

Agricultural Research Council / Medical Research Council: Report of Committee on Food and Nutrition Research (HMSO, London 1974).

American Medical Association, Council on Foods and Nutrition: Iron in enriched wheat flour, farina, bread, buns and rolls. J. Am. med. Ass. *220:* 855–859 (1972).

AMINE, E. K. and HEGSTED, D. M.: Effect of diet on iron absorption in iron-deficient rats. J. Nutr. *101:* 927–935 (1971).

ANDERSON, C. M.: The metabolism of sulphur in the rumen of the sheep. New Zeald. J. Sci. Technol. Sec. A *37:* 279–394 (1956).

ANDREWS, J. W.; MURAL, T., and CAMPBELL, C.: Effects of dietary calcium and phosphorus on growth, food conversion, bone ash and haematocrit levels of catfish. J. Nutr. *103:* 766–771 (1973).

ARMAN, P.; KAY, R. N. B.; GOODALL, E. D., and SHARMAN, G. A. M.: The composition and yield of milk from captive red deer *(Cervus elaphus L)*. J. Reprod. Fert. *37:* 67–84 (1974).

ARNAUD, C. D.; TSAO, H. S., and LITTLEDIKE, T.: Calcium homeostasis, parathyroid hormone and calcitonin. Preliminary Report. Mayo Clin. Proc. *45:* 125–131 (1970).

ASHLEY, L. M.: Nutritional Pathology; in HALVER Fish nutrition, p. 480 (Academic Press, New York 1972).

ASHWORTH, A.; MILNER, R. F.; WATERLOW, J. C., and WALKER, R. B.: Absorption of iron from maize *(Zea mays L)* and soya beans *(Gycine hispida* Max.) in Jamaican infants. Br. J. Nutr. *29:* 269–278 (1973).

AXELROD, D. R. and BASS, D. E.: Electrolytes and acid-base balance in hypothermia (1972). Am. J. Physiol. *186:* 31–34 (1956).

BASHIRULLAH, A. K. M. and ANAM, K.: Calcium and phosphorus contents of bone of some marine and fresh water fishes. Scientific Res., University of Dacca *7:* 83–86 (1970).

BAUER, G. C. H.; CARLSSON, A., and LINDQUIST, B.: Metabolism and homeostatic function of bone; in COMAR and BRONNER Mineral metabolism, vol. 1 B, pp. 609–676 (Academic Press, New York 1961).

BELL, G. H.; CUTHBERTSON, D. P., and ORR, J.: Strength and size of bone in relation to calcium intake. J. Physiol., Lond. *100:* 299–317 (1941).

BERRIDGE, M. J. and PATEL, N. G.: Insect salivary glands: stimulation of fluid secretion by 5-hydroxytryptamine and adenosine 3′, 5′-monophosphate. Science *162:* 462–463 (1968).

BIDDULPH, D. M.; HIRSCH, P. F.; COOPER, C. W., and MUNSON, P. L.: Effect of thyroparathyrodectomy and parathyroid hormone on urinary excretion of calcium and phosphate in the golden hamster. Endocrinology. *87:* 1346–1350 (1970).

BILBY, L. W. and WIDDOWSON, E. M.: Chemical composition of growth in nestling blackbirds and thrushes. Br. J. Nutr. *25:* 127–134 (1971).

BLAXTER, K. L.; ROOK, J. A. F., and MACDONALD, A. M.: Experimental magnesium deficiency in calves, clinical and pathological observations. J. comp. Path. Ther. *64:* 157–175 (1954a).

BLAXTER, K.L. and ROOK, J.A.F.: Metabolism of calcium, magnesium and nitrogen and magnesium requirements. J. comp. Path. Ther. *64:* 176–186 (1954b).

BONHAM, K.: Growth rate of giant clam *Tridacna cigas* at Bikini atoll as revealed by radioautography. Science *149:* 300–302 (1965).

BORLE, A.B.: Calcium and phosphate metabolism. Annu. Rev. Physiol. *36:* 361–390 (1974).

BOTHWELL, T.H. and FINCH, C.A.: Iron metabolism (Little, Brown, Boston 1962).

BOUTWELL, R.K.; GEYER, R.P.; HALVERSON, A.W., and HART, E.B.: The availability of wheat bran phosphorus for the rat. J. Nutr. *31:* 193–202 (1946).

BOYLE, I.T.; GRAY, R.W.; OMDAHL, J.L., and LUCA, H.F. DE: The Mechanism of adaptation of intestinal calcium absorption to low dietary calcium. J. Lab. clin. Med. *78:* 813–361 (1971).

BRYAN, J.E. and LARKIN, P.A.: Food specifialization by individual trout. J. Fish. Res. Board Can. *29:* 1615–1624 (1972).

BUNGE, G.: Z. Biol. *10:* 326 (1874); cit. BRODY Bioenergetics and growth, p. 804 (Reinhold, New York 1945).

CADDELL, J.L. and GODDARD, D.R.: Studies in protein-calorie malnutrition. I. Chemical evidence for magnesium deficiency. New Engl. J. Med. *276:* 533–535 (1967).

CARE, A.D. and BATES, R.F.L.: Control of secretion of parathyroid hormone and calcitonin in mammals and birds. Gen. comp. Endocr. Suppl. *3:* 448–458 (1972).

CHEEKE, P.R. and AMBERG, J.W.: Comparative excretion by rats and rabbits. J. Anim. Sci. *37:* 450–454 (1973).

CONRAD, M.E. and CROSBY, W.H.: Intestinal mucosal mechanisms controlling Fe absorption. J. clin. Invest. *42:* 926 (1962).

COOK, D.A.: Availability of magnesium: balance studies in rats with various inorganic magnesium salts. J. Nutr. *103:* 1365–1370 (1973).

COONS, C.M.: Iron metabolism. Annu. Rev. Biochem. *33:* 459–480 (1964).

CROSBY, W.H.: The control of iron balance by the intestinal mucosa. Blood *22:* 441–449 (1963).

CUTHBERTSON, D.P.: Mineral and trace element requirements for normal growth and development; in SOMOGYI Nutrition and technology of foods for growing humans, pp. 65–91 (Karger, Basel 1973).

DALL, W.: Food and feeding of some Australian penacid shrimps. Fish Rep. FAO, vol. 2, No. 57, pp. 251–258 (FAO, Rome 1968).

DALL, W.: Osmoregulation in the lobster *Homarus americanus*. J. Fish. Res. Board Can. *27:* 1123–1230 (1970).

DAVIS, P.N.; NORRIS, L.C., and KRATZER, F.H.: Iron utilization and metabolism in the check. J. Nutr. *94:* 407–417 (1968).

DENIS, G.; KUCZERPA, A., and NIKOLAIZUK, N.: Stimulation of bone resorption by increasing dietary protein intake of rats fed diets low in phosphorus and calcium. Canad. J. Physiol. Pharmacol. *51:* 539–548 (1973).

DODD, R.H.: The nutritional requirements of locusts. V. Observations on essential fatty acids, chlorophyll, nutritional salt mixtures, and the protein or amino acid components of synthetic diets. J. Insec. Physiol. *6:* 126–145 (1961).

DORFMAN, A.: In Ciba Symp. Chemistry and Biology of Mucopolysaccharides, p. 90 (J. & A. Churchill, London 1958).

DUFRESNE, L.R. and GITELMAN, H.J.: A possible role of adenyl cyclase in the regulation of parathyroid activity by calcium; in Calcium, parathyroid hormone and the calcitonins, pp. 202–206 (Excerpta Medica, Amsterdam 1972).

DZIEWIATKOWSKI, D.I.: Sulphur; in COMAR and BRONNER An advanced treatise, vol. 2, pp. 175–220 (Academic Press, New York 1962).

EANES, E.D.; HARPER, R.A.; GILLESSEN, I.H., and POSNER, A.S.: An amorphous component in bone mineral. Proc. 4th Eur. Symp. Calcified Tissues, Leiden 1966, p. 24 (Excerpta Medica Foundation, Amsterdam 1966).

ELLIOTT, G.V. and JENKINS, T.M.: Winterfood of trout in three high elevation Sierra Nevada lakes. Calif. Fish Game *58:* 231–237 (1972).

ELWOOD, P.C.: The clinical evaluation of circulating haemoglobin level. Clinics Haemat. *3:* 705–718 (1974).

FITCH, G.D.; HARVILLE, W.E.; DINNING, J.S., and PORTER, F.S.: Iron deficiency in monkeys fed diets containing soybean protein. Proc. Soc. exp. Biol. Med. *116:* 130–133 (1964).

FORBES, E.B.; BLACK, A.; BRAMAN, W.W.; FREAR, D.E.H.; KAHLENBERG, O.J.; MCCLURE, F.J.; SWIFT, R.W., and VORIS, L.: Bull. Pa. Agric. exp. Sta. No. 319 (1935); cit. UK Agricultural Research Council (1965).

FOY, H. and KONDI, A.: Anaemias of the tropics. Relation to iron intake, absorption and losses during growth, pregnancy and lactation. J. trop. Med. Hyg. *60:* 105–118 (1957).

FRIEDEN, E.: The chemical elements of life. Sci. Amer. *227:* 52–60 (1972).

GARABEDIAN, M.; HOLICK, M.F.; LUCA, H.F. DE, and BOYLE, I.T.: Control of 25-hydroxycalciferol metabolism by parathyroid glands. Proc. natn. Acad. Sci. USA *69:* 1673–1676 (1972).

GILLIS, M.B.; EDWARDS, H.M., and YOUNG, R.J.: Studies on the availability of calcium orthophosphates to chickens and turkeys. J. Nutr. *78:* 155–161 (1962).

GLOVER, J. and JACOBS, A.: Observations on iron in the jejunal lumen after a standard meal. Gut *12:* 369–371 (1971).

GUÉGUEN, L. and TRUDELLE, F.: The effect of simultaneous or separate supply of calcium and phosphorus on their utilization by the rabbit. C. r. Acad. Sci. *275:* 1645–1648 (1972).

HEAGY, F.C. and BURTON, A.C.: Effect of intravenous injection of magnesium chloride on the body temperature of the unanaesthetized dog with some observations on the magnesium levels and body temperature in man. Am. J. Physiol. *152:* 407–416 (1948).

HEATON, F.W.: Magnesium requirement for enzymes and hormones. Biochem. Soc. Trans. *1:* 67–70 (1973).

HEGSTED, D.M.: Interactions in nutrition; in MERTZ and CORNTZER Newer trace elements in nutrition, p. 19 (Marcel Dekker, New York 1971).

HELLER, B.I.; HAMMERSTEIN, J.F., and STUTZMAN, F.L.: Concerning the effects of magnesium sulphate on renal function, electrolyte excretion and clearance of magnesium. J. clin. Invest. *32:* 858–861 (1953).

HETZEL, H.R.: Studies on holothurian coelomocytes. Biol. Bull. *125:* 289 (1963).

HEVESY, G.; LOCKNER, D., and SLETTEN, K.: Iron metabolism and erythrocyte formation in fish. Acta physiol. scand. *60:* 256–266 (1964).

HINTZ, H.F. and SCHRYVER, H.F.: Magnesium metabolism in the horse. J. Anim. Sci. *35:* 755–759 (1972a).

HINTZ, H.F. and SCHRYVER, H.F.: Availability to ponies of calcium and phosphorus from various supplements. J. Anim. Sci. *34:* 979–980 (1972b).

HINTZ, H.F.; WILLIAMS, A.J.; ROGOFF, J., and SCHRYVER, H.F.: Availability of phosphorus of wheat bran when fed to ponies. J. Anim. Sci. *36:* 522–525 (1973).

HODGE, H.C. (1938); cit. IRVING, J.T.: Calcium and phosphorus metabolism, p. 129 (Academic Press, New York 1973).

HOLMGREN, H. und WILANDER, O.: Beitrag zur Kenntnis der Chemie und Funktion der Ehrlichschen Mastzellen. Z. mikr.-anat. Forsch. *42:* 242–278 (1937).

HUOVINEN, J.A. and GUSTAFSSON, B.E.: Inorganic sulphate, sulphate and sulphide as sulphur donors in the biosynthesis of sulphur amino acids in germ-free and conventional rats. Biochim. biophys. Acta *136:* 441–447 (1967).

ICHIKAWA, R. and OGURI, M.: Metabolism of radionucleids in fish. I. Strontium-calcium discrimination in gill absorption. Bull. Jap. Soc. Sci. Fish. *27:* 351–356 (1961).

JANDL, J.H. and KATZ, H.J.: The plasma to-cell cycle of transferrin. J. clin. Invest. *42:* 314–326 (1963).

JONES, E.I.; MCCANCE, R.A., and SHACKLETON, L.R.B.: The role of iron and silica in the structure of the radular teeth of certain marine molluscs. J. exp. Biol. *12:* 59–64 (1935).

KAPOOR, U.R.; AGARWALA, O.N.; PACHAURI, V.C.; NATH, K., and NARAYAN, S.: The relationship between diet, the copper and sulphur content of wool and fibre characteristics. J. agric. Sci. *79:* 109–114 (1972).

KATZ, H.J. and JANDL, J.H.: in GROSS Int. Symp. Iron Metabolism (Springer, Berlin 1964).

KAZARIAN, L.E. and GIERKE, H.E. VON: Bone loss as a result of immobilization and chelation. Preliminary results in *Macaca mulatta*. Clin. Orthop. *65:* 67–75 (1969).

KEENE, O.D. and COMBS, G.F.: Magnesium requirement of chicks and poults. Poult. Sci. *41:* 1954 (1962).

KENNEDY, M. and FITZMAURICE, P.: Growth and food of grown trout *Salmo tratta* (L) in Irish waters. Proc. R. Irish Acad. *71B:* 269–352 (1971).

KENNEDY, W.J.; TAYLOR, J.D., and HALL, A.: Environmental and biological controls on bivalve shell mineralogy. Biol. Rev. *44:* 499–530 (1969).

KERR, G.R.: Nutritional problems of subhuman primates. Physiol. Rev. *52:* 415–456 (1972).

KLEIBER, M.; BOELTER, M.D., and GREENBERG, D.M.: Fasting catabolism and food utilization of magnesium deficient rats. J. Nutr. *21:* 363–372 (1941).

KODICEK, E. and LOEWI, G.: Uptake (^{35}S) sulphate by micropolysaccharides of granulated tissue. Proc. R. Soc. B *144:* 100–115 (1955).

KRISHNARAO, G.V.G. and DRAPER, H.H.: Influence of dietary phosphate on bone resorption in senescent mice. J. Nutr. *102:* 1143–1145 (1972).

LEE, D., jr. and MATRONE, G.: Iron and copper effects on ceruloplasmin activity of rats with zinc-induced copper deficiency. Proc. Soc. exp. Biol. Med. *130:* 1190–1194 (1969).

LIM, P. and JACOB, E.: Magnesium status of alcoholic patients. Metab. clin. Exp. *21:* 1045–1051 (1972).

LING, E. R.; KON, S. K., and PORTER, J. W. G.: The composition of milk and the nutritive value of its components; in KON and COWIE The mammary gland and its secretion, pp. 195–263 (Academic Press, New York 1961).

LOEWI, G. and KENT, P.W.: The utilization of inorganic sulphate by granulation tissue. Biochem. J. *65:* 550–554 (1957).

LOWE, I.P. and ROBERTS, E.: Incorporation of radioactive sulphate sulphur into taurine and other substances in the chick embryo. J. biol. Chem. *212:* 477–485 (1955).

LOWENHAUPT, E.; SCHULMAN, M.P., and GREENBERG, D.M.: Basic histologic lesions of magnesium deficiency in the rat. Archs. Path. *49:* 427–433 (1950).

LYMAN, C.P. and CHATFIELD, P.O.: Physiology of hibernation in mammals. Physiol. Rev. *35:* 403–425 (1955).

MACHLIN, L.J.; PEARSON, P.B., and DENTON, C.A.: The utilization of sulphate sulphur for the synthesis of taurine in the developing chick embryo. J. biol. Chem. *212:* 469–475 (1955).

MCCANCE, R.A. and WIDDOWSON, E.M.: The absorption and excretion of iron following oral and intravenous administration. J. Physiol., Lond. *94:* 148–154 (1938).

MARSHALL, J.H.: Measurements and models of skeletal metabolism; in COMAR and BONNER Mineral metabolism: an advanced treatise. Calcium physiology, vol. 3, pp. 12–122 (Academic Press, New York 1969).

MAZUR, A.; GREEN, S., and CARLETON, A.: Mechanism of plasma iron incorporation into hepatic ferritin. J. biol. Chem. *235:* 595–603 (1960).

MILLS, C.F.: The detection of trace element deficiency and excess in man and farm animals. Proc. Nutr. Soc. (in press, 1975).

MORGAN, R.S. and DAVIS, H.P.: The effect of pregnancy and parturition on the weight of dairy cows. Res. Bull. Nebr. Agric. exp. Sta., No. 82 (1936).

MORGULIS, S.: Studies on the chemical composition of bone ash. J. biol. Chem. *93:* 455–466 (1931).

MÜLLER-REH, F.: The mineral (Ca, P, Mg, K, Na) and trace element (Fe, Cu, Zn, Mn) supply in the horse; thesis Tierärztliche Hochschule, Hannover (1972).

MUNSON, P.L.; TASHYIAN, A.H., jr., and LEVINE, L.: Evidence for parathyroid hormone in non-parathyroid tumors associated with hypercalcaemia. Cancer Res. *25:* 1062–1067 (1965).

MURRAY, M.J. and STEIN, N.: Experimental achylia gastrica and iron absorption. Br. J. Haemat. *21:* 113–120 (1971).

NICOL, J.A.C.: The biology of marine animals (J. Wiley & Sons, Chichester 1969).

NORRIS, B.A. and TAYLOR, T.G.: The daily food consumption of laying hens in relation to egg formation. Br. Poult. Sci. *8:* 251–257 (1967).

O'BRIEN, J.R.P.: Absorption and storage of iron. Biochem. Soc. Trans. *1:* 70–73 (1973).

O'DELL, B.L. and CAMPBELL, B.J.: Trace elements: metabolism and metabolic function; in Metabolism of vitamins and trace elements, vol. 21, pp. 179–265 (Elsevier, Amsterdam 1971).

OGASAWARA, K.: Total and ionized magnesium in serum. Igaku to Seibutsugakn *29:* 250 (1953); cit. WACKER and VALLEE (1974).

O'KELLEY, R.E. and FONTENOT, J.P.: Effects of feeding different magnesium levels to dry lot-fed lactating beef cows. J. Anim. Sci. *29:* 959–966 (1969).

OLSSON, N.O.: The nutrition of the horse; in CUTHBERTSON Nutrition of animals of agricultural importance. Part 2. Assessment of and factors affecting requirements of farm livestock, in SINCLAIR Encyclopaedia of food and nutrition, vol. 19, pp. 921–960 (Pergamon, Oxford 1969).

ORME-JOHNSON, W. H.: Iron sulphur proteins: structure and formation. Annu. Rev. Biochem. *42:* 159–204 (1973).

PHILLIPS, A. M., jr.; PODOBAK, H. A.; POSTON, H. A.; LIVINGSTONE, D. L.; BOOKE, H. E.; PYLE, E. A., and HAMMER, G. L.: N. Y. Cons. Dept. Cort and Hatchery Rep. 32 (1964).

PLA, G. W. and FRITZ, J. C.: Collaborative study of the haemoglobin repletion test in chicks and rats for measuring availability of iron. J. Ass. Off. Anal. Chem. *54:* 13–17 (1971).

PLATNER, W. S.: Effects of low temperature on magnesium content of blood, body fluids and tissues of goldfish and turtle. Am. J. Physiol. *161:* 399–405 (1950).

PLATNER, W. S. and HOSKO, M. G.: Mobility of serum Magnesium in Hypothermia. Am. J. Physiol. *174:* 273–276 (1953).

POSNER, A. S.: Crystal chemistry of bone mineral. Physiol. Rev. *49:* 760–792 (1969).

PRENTICE, J. and MATRONE, G.: Molybdenum inhibition of sulphate reduction in rumen microorganisms. Fed. Proc. Fed. Am. Socs exp. Biol. *29:* 766 (1970).

PYKE, R. E.; MACK, P. B.; HOFFMAN, R. A.; GILCHRIST, W. W.; HOOD, W. N., and GEORGE, G. P.: Physiologic and metabolic changes in *Macaca nemestrina* on two types of diet during restraint and non-restraint. III. Excretion of calcium and phosphorus. Aerospace Med. *39:* 704–708 (1968).

RANHOTRA, G. S.; HEPBURN, F. N., and BRADLEY, W. W.: Availability of iron in enriched bread. Cereal Chem. *48:* 377–384 (1971).

RATH, C. E. and FINCH, C. A.: Chemical, clinical and immunological studies on the products of human plasma fractionation. XXXVIII. Serum iron transport. Measurement of iron binding capacity of serum in man. J. clin. Invest. *28:* 79–85 (1949).

REED. L. J.; CAVALLINI, D.; PLUM, F.; RACHELE, J. R., and VIGNEAUD, V. DU: The conversion of methionine to cystine in a cystinuric. J. biol. Chem. *180:* 783–790 (1949).

REEN, R. VAN and PEARSON, P. B.: Magnesium deficiency in the duck. J. Nutr. *51:* 191–203 (1953).

REINHOLD, J. C.; LAHIMGARZADEH, A.; NASR, K., and HEDAYATI, H.: Effects of purified phytate and phytate-rich bread upon metabolism of zinc, calcium, phosphorus and nitrogen in man. Lancet *ii:* 22 (1973).

RICHELLE, L. J. and ONKELINX, C.: Recent advances in the physical biology of bone and other hard tissues; in COMAR and BRONNER Mineral metabolism – an advanced treatise, vol. 3, pp. 123–190 (Academic Press, New York 1969).

RIEDESEL, M. L. and FOLK, G. E.: Serum magnesium and hibernation. Fed. Proc. Fed. Am. Socs exp. Biol. *15:* 157 (1956).

ROBERTSON, J. D.: Ionic regulation in some marine invertebrates. J. exp. Biol. *26:* 182–200 (1949).

ROSEN, E. V.; CAMPBELL, P. G., and MOOSA, G. M.: Hypomagnesaemia and magnesium therapy in protein-calorie malnutrition. J. Pediat. *77:* 709–714 (1970).

ROTHMAN, R. H.; KLEMEK, J. S., and TOTON, J. J.: The effect of iron deficiency anaemia on fracture healing. Clin. Orthop. *77:* 276–283 (1971).

RUUD, J.T.: Vertebrates without erythrocytes and blood pigment. Nature, Lond. *173:* 848–850 (1954).

SAMPSON, H.W.; MATTHEWS, J.L.; MARTIN, J.H., and KUNIN, A.S.: An electron microscopic localization of calcium in the small intestine of normal rachitic and vitamin D_3-treated rats. Calcif. Tissue Res. *5:* 305–316 (1970).

SANFORD, R.: Release of iron from conjugates in foods. Nature, Lond. *185:* 533–534 (1960).

SANG, J.H.: The quantitative nutritional requirements of *Drosophila melanogaster*. J. exp. Biol. *33:* 45–72 (1956).

SCHMIDT, J.J.; BRUSCHKE, G.; GROSSMANN, I.; KALBE, I.; DRANZ, D.; LINDERHAYN, K.; MEHLS, E., and REITZIG, P.: The occurrence of osteoporosis in iron deficiency. Dt. Gesundh. Wes. *26:* 2348 (1971).

SCOTT, M.L.; BUTTERS, H.E., and RANIT, G.O.: Studies on the requirements of young poults for available phosphorus. J. Nutr. *78:* 223–230 (1962).

SEKI, H.: The effect of calcium level of diet on the metabolism of calcium, phosphorus and magnesium in pregnant rats. 2. Mineral contents of foetuses. J. Nutr. *29:* 129–132 (1971).

SHOHL, A.T.: Mineral metabolism. Am. Chem. Soc. Monogr. Ser. (Reinhold, New York 1939).

SHTENBERG, A.I. and TORCHINSKII, A.M.: Composition of a mineral mixture for an artificial diet for rats. Vestnik Akademiya Meditsinskikh Nauk SSR *27:* 47–52 (1972); cit. Nutr. Abstr. Rev. *43:* 568–569 (1973).

SMELOVA, I.V.: Absorption of S^{35}-labeled compounds from water by fish. Chem. Abstr. *57:* 1394c (1962).

SMITH, R.H.: Calcium and magnesium metabolism in calves. Biochem. J. *67:* 472–481 (1957).

SOUTHWARD, E.C.: Pogonophora. Oceanogr. Mar. Biol. Ann. Rev. *1:* 405 (1963).

SPITZER, R.R.; MARUYAMA, G.; MICHAUD, L., and PHILLIPS, P.H.: The role of vitamin D in the utilization of phytin phosphorus. J. Nutr. *35:* 185–193 (1948).

SULLIVAN, T.W.: Magnesium requirement of turkeys to 4 weeks of age. Poult. Sci. *41:* 1686–1687 (1962).

SWAN, C.H.J. and COOKE, W.T.: Nutritional osteomalacia in immigrants in an urban community. Lancet *ii:* 456–459 (1971).

TALMAGE, R.V.; WHITEHURST, L.A., and ANDERSON, J.J.B.: Effect of calcitonin and calcium infusion on plasma phosphate. Endocrinology *92:* 792–798 (1973).

TEMPLETON, W.L. and BROWN, V.M.: Accumulation of calcium and strontium by brown trout from waters in the United Kingdom. Nature, Lond. *198:* 198–200 (1963).

UK Agricultural Research Council: The nutrient requirements of farm live stock. No. 1 Poultry – Technical Reviews and Summaries (ARC, London, 1975).
No. 2 Ruminants – Technical Reviews and Summaries (ARC, London 1965).
No. 3 Pigs – Technical Reviews and Summaries (ARC, London 1967).

UK Department of Health and Social Security: Recommended intakes of nutrients for the United Kingdom. Rep. publ. Hlth and Med. Subj., No. 120 (HMSO, London 1969).

ULLREY, D.E.; ZUTAUT, C.L.; BALTZER, B.V.; VINCENT, B.H.; SCHMIDT, D.A.; HOEFER, J.A., and LUECKE, R.W.: Magnesium requirement of the baby pig. J. Anim. Sci. *21:* 1006–1007 (1962).

UPTON, A.C. and ODELL, T.T.: Utilization uf ^{35}S-labeled sulphate in scorbutic guinea pigs. Archs. Path. *62:* 194–199 (1956).

USA Academy of Sciences, National Research Council, Committee on Animal Nutrition: Nutrient requirements of domestic animals. 6. Nutrient requirements of horses. Publ. No. 912 Washington, D. C. (revised) (1961). 8. Nutrient requirements of dogs. Publ. No. 989, Washington, D. C. (1962), 10. Nutrient requirements of Laboratory Animals. Publ. No. 990. Ed. 2 (revised). Washington, D. C. (1962). 11. Nutrient requirement of trout, salmon and catfish. Washington, D. C. (1973). Recommended dietary allowances, 8th ed., Washington, D. C. (1974).

USA Food and Drug Administration: Recommended daily allowances. Washington, D. C. (1973).

VILLEE, C.A. and DETHIER, V.G.: Biological principles and processes, p. 87 (Saunders, Philadelphia 1972).

VITALE, T.T.; VELEZ, H.; GORZMAN, C., and CORREA, P.: Magnesium deficiency in the Cebus monkey. Circulation Res. *12:* 642–650 (1963).

WACKER, W.E.C. and VALLEE, B.L.: Magnesium; in COMAR and BRONNER Mineral metabolism, an advanced treatise. The elements, vol. 2 (Academic Press, New York 1974).

WADDEL, D.G. and SELL, J.L.: Effects of dietary calcium and phosphorus on the utilization of dietary iron by the chick. Poult. Sci. *43:* 1249–1257 (1964).

WALKER, A.R.P.; FOX, F.W., and IRVING, J.T.: Studies in human mineral metabolism. I. The effect of bread rich in phytate phosphorus on the metabolism of certain mineral salts with special reference to calcium. Biochem. J. *42:* 452–462 (1948).

WHANGER, P.D.: Sulphur in the ruminant. Wld Rev. Nutr. Diet., vol 15, pp. 225–255 (Karger, Basel 1972).

WILLS, M.R.; DAY, R.C.; PHILLIPS, J.B.; BATEMAN, E.C.; WILSON, H.R.; HOLLAND, M.W., and HARMS, R.H.: Dietary calcium and phosphorus requirements for bob-white chicks. J. Wildlife Mgmt *26:* 965–968 (1972).

WINEGRAD, S.: Calcium and striated muscle; in COMAR and BRONNER Mineral metabolism, an advanced treatise, vol. 3, pp. 191–233 (Academic Press, New York 1969).

Sir DAVID CUTHBERTSON, University Department of Pathological Biochemistry, Royal Infirmary, *Glasgow* (Scotland)

Comp. Anim. Nutr., vol. 2, pp. 87–143 (Karger, Basel 1977)

Trace Elements in Animals

R. B. WILLIAMS
Department of Nutritional Biochemistry, Rowett Research Institute, Bucksburn, Aberdeen

I. General Considerations

A. Nomenclature

It has been recognised for many years that all living organisms contain only minute amounts of certain inorganic elements. Initial difficulties in the accurate determination of the low concentrations of many of these elements led inevitably to their description as *'trace elements'*. Modern analytical techniques now permit the swift and accurate analysis of these substances with a consequent explosion of interest in their metabolism and function. Regrettably, however, the initial distinction between 'trace' and macro-elements (such as Ca, Mg, Na and K) persists, particularly in the minds of those studying vertebrate nutrition, with the result that the impression is given that classes of elements exist which may have quite separate and distinctive biological roles. Such arbitrary distinction takes no account of the fact that in soft tissues of animals the concentration of an element that may be regarded by some as, nominally, a 'trace' element may even exceed that of certain of the macro-elements. In invertebrate animals the overall concentration of specific elements of either class may not be substantially different.

With this proviso in mind, the present article will, for reasons of brevity, be concerned with those elements generally accepted as essential 'trace elements' which now include Cr, Co, Cu, F, I, Mn, Mo, Ni, Se, Si, Sn, V and Zn. No specific consideration will be given to Fe in this text.

A rigid criterion of the *essentiality* of a given element must include the demonstration that the absence of the element adversely affects the biochemical or metabolic processes of an organism, and that restoration of the defect is *specific* to that element. By this criterion it is not yet possible to claim that all the above-mentioned elements are essential for all forms of animal life. In the great majority of instances it can only be assumed that similar metabolic pathways in differing species may be dependent on the same mineral elements.

The demonstration of a physiological requirement for an inorganic nutrient in any animal species depends on the ability of that species to grow and reproduce while assimilating a diet the elemental composition of which can be controlled under closely defined experimental conditions. This clearly precludes experimentation on species of fastidious eaters, symbionts and many classes of parasites. For this, and other reasons (many economic), progress in elucidation of trace element requirements and function has, in the main, been confined to studies on man and domestic animals.

Definition of basic requirements for specific physiological processes rests upon the demonstration that, in the absence of complicating factors which may affect the availability of the test element, the addition of a mineral element in graded amounts to a test diet provokes a physiological response which reaches an optimum at a given concentration of that element. It is convenient, therefore, to consider *functional requirements* in terms of dietary *concentration* and usually in mg/kg of the diet.

The definition of the biochemical significance of the element may reside in showing that the element is involved in: (a) the functioning of an enzyme; (b) the stabilisation of macromolecular structures of the cell, or (c) is otherwise concerned in the control of metabolic events. These properties are not necessarily mutually exclusive, and may not be shared by all elements.

Such characteristics suggest that essential trace elements function by virtue of their frequent association with organic molecules and that the free ionic form of the element may be only a transitory state indicative of the flux of that element from one tissue component to another. The simplest complexes of element and organic molecule probably occur in the gut or surrounding fluid and exist chiefly as chelates of amino acids, peptides and, possibly, porphyrins. When absorption has occurred, competition for the available element will depend on the relative stabilities of the metal-ion complexes in the cell. Within a cell, many such complexes may be found, so that a deficiency of the element in question may not be expected to affect all processes dependent on that element to the same degree. In higher animals where considerable differentiation of cell types has taken place, a differential response in the susceptibility of various organs to a trace element deficiency may be expected. Differences in stability of the complexes formed between metal ions and their organic ligands are exemplified by the distinctions between *metal-enzyme complexes* and *metalloenzymes* as tabulated by PARISI and VALLEE [247]. These may be briefly summarised as follows: A firm binding of metal to protein in the case of metalloenzymes, together with retention of metal during purification and with an approach to a fixed stoichometric limit, as opposed to a loss of metal from the weaker metal-enzyme complex. Proportionality also exists between metal content and enzymic activity in metalloenzymes such that no further activation is possible when the full metal complement is reached. This is in contrast to the susceptibility of the metal-enzyme complex to 'activation' by the addition of a variety of metal ions. Further, in metalloenzymes it is often possible to assign a specific biological role to the metal.

The occurrence of deficiency states in the animal may arise as a result of a *simple* deficiency when insufficient amounts of the essential element are present to effect saturation of the functionally important binding sites within the organism. A *conditioned* deficiency occurs most frequently when, in the presence of a normally adequate amount of the required element in the food or external environment, the simultaneous presence of dietary components having a high affinity for the element will limit its uptake into carrier systems or restrict its incorporation into functionally active sites. Conditioned deficiency also arises when, by virtue of analogies in atomic structure, two or more elements may compete for such sites. Several examples of these types of conditioned deficiency will be described later in this text.

B. Occurrence

Except for the abundant Si, the average concentration in the earth's crust of those trace elements essential to life ranges from about 1,000 g/tonne in the case of Mn to 0.1 g/tonne for Se. Within this wide range occur localised geographical variations, in a number of instances sufficiently large to produce deficiencies or excesses in the amount of mineral available to both flora and fauna of those areas. As within intracellular systems, the presence or absence of natural or synthetic chelators may influence the response of whole populations to the trace element content of their environment. Such an effect in the marine environment is illustrated by HUTNER [147] who cites examples of the induction of blooms of diatoms in the waters of the North and Sargasso seas by the addition of the chelating agent EDTA (ethylenediaminetetra-acetic acid) which increased the availability of essential minerals to these organisms. In the case of the Sargasso sea, Fe occurs almost entirely in the form of unavailable particulate ferric complexes. The occurrence, therefore, of an element in the environment cannot be regarded as an index of its adequacy or potential toxicity.

The essentiality of most elements for many species must remain unconfirmed. As pointed out by PARISI and VALLEE, phylogenetic comparisons may suggest useful lines of research, but cannot be interpreted as implying that, in differing species, mechanisms do not exist which may be functional in the absence of one or other essential minerals.

The study of trace elements in the animal body has been carried out most intensively in domestic and laboratory animals, and only in these cases can any comparative assessment of mineral status be made. 'Normal' values for the concentration of a particular element may be influenced by age, sex, pregnancy or genetic considerations. Disease states may also modify tissue metal distribution [249, 262]. The effects of nutritional status, pregnancy, disease state and environmental factors on the trace element content of blood and of milk of various animal species has been adequately described by UNDERWOOD [351] and it would be inappropriate to attempt to summarise that information.

With some exceptions, mineral element concentrations in structural organelles of tissues are less susceptible to dietary changes than are those of body fluids. Species differences in metal accumulation and distribution are, at present, somewhat eclipsed by the many reports on the readiness with which certain organisms, notably among the mollusca, may accumulate large amounts of heavy metals. However, pigs, ducks, sheep and young cattle have been known for a long time for their ability to store considerable quantities of Cu in the liver. If these peculiarities are ignored, it can be assumed that in many species normal adults may have on average between 10 and 40 μg Cu/g wet weight of liver, and a mean body concentration of about 2 μg/g. The new-born young of many species contain two to three times as much Cu per unit of body weight as do adults, and a great part of this Cu is deposited in the liver. The distribution of Cu within the animal body has been studied by CUNNINGHAM [59] and by SCHROEDER *et al.* [298].

Cobalt, unlike Cu, does not accumulate in foetal liver; neither is there evidence for abnormally high concentrations in any particular tissue. In nearly all tissues Co concentration is less than 0.1 μg/g wet weight. The distribution of Co in many species has been studied [31, 176, 177, 277].

Analytical difficulties in the determination of Cr may be responsible for wide variations reported in the distribution and concentration of this element in animal tissues. The possibility that the biologically active form of Cr exists in combination with an organic ligand which, under some circumstances, may be volatile means that highly specific techniques are necessary for its accurate determination. Little data is yet available on the Cr content of animal tissues. The studies of TIPTON [348] show that infant human tissues have a higher Cr concentration than do those of adults; the latter usually contain approximately 0.03 μg/g on a dry matter basis.

In bones and teeth, unlike soft tissues, high levels of F occur. A number of examples quoted by UNDERWOOD [351] suggest that F levels in bone of animals may normally be in the range of 300–600 μg/g of fat-free dry matter, with slightly lower values in teeth. Soft tissues, in contrast, contain about 2–4 μg/g dry matter. When F intakes are high, concentrations in bone may rise to as high as 20,000 μg/g [342], with very much smaller increases in the soft tissues.

Iodine accumulation in the thyroid gland (0.2–0.5% I on dry matter basis) accounts for 70–80% of the total body I in the adult human [351]. In other tissues, total I concentration (inorganic I plus organic I) probably does not exceed 0.05 μg/g, although higher levels have been found in the ovaries and salivary and pituitary glands.

Manganese concentrations in animal tissues are low. In most tissues, less than 1 μg/g wet weight may be found [98–100, 296] although concentrations in kidney, liver, pancreas and pituitary gland lie between 1 and 3 μg/g, and bones contain more than 3 μg/g. In hair, wool and feathers Mn concentrations appear to reflect the adequacy or otherwise of dietary Mn [351]. Variations in Mn content of hen eggs according to dietary Mn supply have also been reported [135].

Like Mn, tissue concentrations of Mo are generally low. UNDERWOOD [351] cites values for normal animals ranging from 0.06 μg Mo/g dry matter for muscle to 4.4 μg/g in the case of kidney, although increases in the concentration of Mo in bones, kidney, liver, skin, hair and wool may occur under conditions of high Mo intake.

The occurrence of Ni in human and animal tissues has been extensively reviewed by NIELSEN [237]. Although no specific biochemical role for this element has yet been established in animal tissues, the occurrence of Ni in above average concentrations in certain tissues, especially brain and aorta, suggests that it may have a role in the control of cellular events. Nickel concentrations in human tissues [as cited by NIELSEN] range from 0.002 to 0.023 μg Ni/ml plasma to 6 μg/g in aorta and about 110 μg/g bone ash.

Selenium is widely distributed in low concentration throughout the animal body, with relatively high levels in kidney, this organ being particularly sensitive to changes in dietary Se levels. Se levels in human tissues were reported by DICKSON and TOMLINSON [73] who found mean values of about 0.4 μg Se/g. Similar values have been reported in pigs [186]. In rats, BURK *et al.* [39] found a mean carcase concentration of 0.2 μg Se/g.

Silicon distribution in animal tissues has been studied by KING and BELT [167]. Lower concentrations occur in foetal than in adult tissues, with ranges from 40–400 μg/g dry weight to 50–1,000 μg/g, respectively, being reported. Other values, cited by UNDERWOOD [351], range from 18 μg/g dry weight for human muscle to 106 μg/g for epidermis.

Earlier statements regarding the occurrence and distribution of Sn as determined by conventional methods have been criticised by SCHWARTZ [302], who has pointed out the relative ease with which organo-tin compounds may be lost from biological samples by evaporation at temperatures well below those needed for destruction of organic matter. Mean values for Sn in various human tissues reported by SCHROEDER *et al.* [295] ranged from 0.23 to 1.20 μg/g wet weight. KEHOE *et al.* [160] in an earlier study reported values ranging from 0.1 μg/g in muscle to 0.8 μg/g in fresh bone.

The concentration of V in animal tissues other than ascidian worms [181], appears to be very low. SÖREMÄRK [328], using a neutron activation technique, has reported lower concentrations of V in a variety of biological specimens than have other workers, although extreme variability in V concentration appears in and between species. Reported values range from undetectable amounts in various organs [328] to rather more than 12 μg/g in fats and oils [294]. Significantly higher values for lung as compared with other soft tissues were obtained by SCHROEDER *et al.* [294].

One of the most intensively studied elements in recent years is Zn. Even so, a certain disagreement is evident in the literature regarding the content and distribution of Zn within the animal, and this reflects, as in the case of other elements, differences in operational techniques, dietary standards and experimental protocols. No doubt exists, however, that marked differences occur in the concentration of Zn in various tissues. High levels of Zn occur in bone, teeth, hair and wool. SPRAY and WIDDOWSON [332, 363] estimate that 38% of the whole body Zn of rat and hedgehog is contained in skin and hair or skin and bristles. The prostate gland may contain 100–230 μg Zn/g fresh tissue depending on species, a concentration only marginally exceeded by bone, but well below the astonishing concentration of 138,000 μg/g dry weight in the tapetum lucidum of the fox eye [351]. Other soft tissues generally contain from approximately 10–50 μg Zn/g fresh weight according to species, although muscle Zn content appears to depend on colour and activity [45]. Differences in Zn content of fast and slow-acting muscles of the lobster *(Homarus vulgaris)* were demonstrated by BRYAN [36]. In hen eggs, about 1 mg of Zn occurs, almost exclusively in the yolk [25]. Overall body Zn concentrations usually lie between 20 and 30 mg/kg fresh weight according to species, with somewhat lower levels in the young animal. Comprehensive values for Zn concentration in many animal and plant tissues have been published by SCHROEDER *et al.* [299].

II. Qualitative Requirements of Different Animals

A. Invertebrates

1. Protozoa

The inorganic nutrition of the many classes of protozoa is ill-defined. The inclusion in culture media of salts of Fe, Mn, Zn and Cu is commonplace and appears to be the result of arbitrary decisions that a probable requirement exists. In some instances this may have been the result of an observation that growth in a 'pure' nutrient solution may be stimulated by a solution of the ash of casein, serum or other plant or animal material. Spectrographic analysis of the ash may reveal the existence of a wide range of elements, not all of which may be essential, while others, such as Se, I and Sn may be wholly or partly lost during ashing procedures. The selection of those elements regarded as probably essential appears therefore to rest on the demonstrated requirements of metazoa.

Inorganic nutrition in microbiology has recently been reviewed by HUTNER [147]. Of the many important observations made by this author the necessity for providing mixes containing trace elements in adequate quantities and in correct proportions with use of a chelating agent, such as citric acid or NTA (nitrilotriacetic acid), warrants particular attention. As has already been mentioned, the metazoan cell exists in a milieu of organic constituents of varying complexity to which are bound, with differing orders of stability, those elements essential, or otherwise, to life processes. There is, at present, no evidence to suggest that the protozoan cell is any more able to accomodate to the presence of uncomplexed trace elements.

As pointed out by HUTNER *et al.* [149], the identification of essential trace elements is difficult except when the production or fate of a key metabolite is associated with the element in question. An attempt to study the inorganic requirements of *Tetrahymena* was made by KIDDER *et al.* [164]. The organisms were grown in a pure (bacteria-free) culture. No requirement for Mn, Zn, Co, F, B or Mo was demonstrable, and the authors pointed out that a previously determined requirement for Cu for this organism was eliminated by a change in the source of amino acids used in the medium, suggesting that Cu may have occurred as a contaminant in the later work. This is illustrative of the difficulties inherent in attempting to show that microorganisms have specific requirements for trace elements particularly when, as a result of the disparity between the mass of organism and that of the culture medium, even minor contamination becomes important.

The effects of elevated incubation temperatures in modifying the nutrient requirements of microorganisms has been applied by HUTNER *et al.* [148] to study the effect of temperature on the demand by those organisms for increased trace element levels in the culture medium. *Ochromonas malhamensis* required additional Zn, Mn and Cu as incubation temperature was raised from 34 to 35.9 or 36.3°C. The possibility that changes in trace element requirement may be an important component of the response of lower organisms to temperature changes is indicated also by work with *Euglena gracilis* and *Crithidia fasciculata.* HUTNER suggests that the imposition of stress upon an organism may induce a condition when one or more trace elements can be limiting.

Circumstantial evidence for trace element requirements in protozoa depends on the identification of enzyme systems which, in other species, have been shown to be dependent on the presence of metals. The occurrence of alkaline phosphatases and lactic and malic dehydrogenases in many species studied suggest a requirement for Zn; cytochrome oxidase for Cu and isocitric dehydrogenase for Mn. Conversely, the absence of xanthine oxidase from *Tetrahymena* possibly indicates that no requirement exists for Mo in this organism.

Clearly, a considerable amount of work remains to be done to identify trace element requirements in simple organisms. The difficulties involved in the selection and purification of media must not be underestimated; unfortunately, they are often completely ignored! An outstanding exception to this generalisation is a study on some effects of Zn deficiency on metabolism in *Euglena gracilis* [265].

2. Insects

Insect nutrition has recently been reviewed by DADD [62] who pointed out that the difficulties in devising suitable diets in which trace mineral content could be controlled are reflected by the present lack of knowledge of trace element requirements of insects. A comprehensive bibliography of insect diets compiled by SINGH [320] covering the period 1900–1970 cites 980 references, about 2% of which refer to inorganic nutrition. However, several trace metals (Fe, Zn, Mn and Cu) have been shown to be essential for various species which include the aphids *Myzus persicae, Apis fabia, Neomyzus circumflexus* and the larvae of *Bombyx mori* and *Tribolium confusum.* Zn was also found necessary for *Tenebrio molitor* [101]. The identification of Cu, Mn and Zn as essential metals for the cockroach *Blatella germanica* (Linnaeus) [114] was followed by the observation of BROOKS [35] that Zn acted as synergist to Mg for the normal transmission in this species of symbiotic bacteroid to following generations. Recent work by AUCLAIR and SRIVASTAVA [14] suggest that in the pea aphid, *Acyrthosiphon pisum,* a requirement for B, Mo and Co may also exist. These authors also showed that the tolerance of *A. pisum* to all the metals tested was fairly high, but that toxicity symptoms appeared when Na borate and Na molybdate reached levels of about 50 mg/100 ml diet.

AKEY and BECK [2] successfully reared *A. pisum* for more than 46 generations on sterile holidic diets. Their work demonstrated that Fe deficiency severely depressed performance within one generation and within 2 generations a deficiency of Cu became operative. Reproduction was more severely limited by trace mineral deficiency than was growth. Fe, Zn, Cu and Mn were all required. *Schizaphis graminum* required Zn, Fe and Mn in the diets of CRESS and CHADA [58] before optimum growth and development occurred when the performance was equivalent to that on a susceptible strain of barley.

The presence in the diet of trace metals in sequestered form improved the performance of *A. pisum* [335]. The ability of different aphid species to accomodate to minerals supplied as metal salts or as chelated complexes is referred to by DADD [62] who showed that, in the case of *M. persicae* maintained on synthetic diets, the presence of citric or ascorbic acids as chelators in these diets markedly improved performance.

Interactions between trace metals were studied by MEDICI and TAYLOR [205] in one of the most comprehensive examinations of insect mineral requirements. Working with the confused flour beetle *(Tribolium confusum)* they established a requirement for Fe, Zn and Mn but not for Cu in excess of the basal 0.12 μg/g diet. The presence of Cd was shown to increase the requirement for Zn by this species [206] and, in addition, it was established that a relationship existed between toxic levels of Zn and dietary Cu levels.

The addition of Fe, Mn, Zn and Cu to honeybee *(Apis mellifica)* diets was suggested by NATION and ROBINSON [233]. Analysis of adult bees for trace metals showed a high concentration of Mn (102–207 μg/g) and this element as well as Cu and Zn was found to be concentrated in the abdomen. LANG [172] examined the Zn content of 7 insect species *(Aedes aegypti, Culex pipiens, Eulex molestus, Tribolium confusum, Oncopeltus fasciatus, Lymantia dispar* and *Apis mellifera* – representing 5 biological orders) and concluded that the Zn content of most species ranged between 0.05 and 0.2 μg/mg dry weight. Higher concentrations of Zn were found in mosquito species and this element was again concentrated in the abdomen.

The metabolic role of metals in most species appears not to have been studied. The identification of metalloenzymes in insect species is also limited, although in recent work, ASHIDA [12] has shown that a pre-phenol-oxidase extracted from the haemolymph of *Bombyx mori* larvae contained about 0.15% of Cu and 2 SH groups per atom Cu. That an alkaline phosphatase isolated from *Drosophila melanogaster* required Zn for activation was shown by HARPER and ARMSTRONG [123].

General conclusions are that Fe, Mn, Cu, Zn are required by all insects and that B, Co and Mo may also be essential. It should, however, be pointed out that requirements for mineral elements determined in larval stages of insect development may represent to some extent the laying down of stores necessary for reproduction in the adult or for metamorphosis. In this connexion LANG suggested that the high levels of Zn found in mosquito abdomen were probably related to the high reproductive capacity of this insect, and it is probable that for reproductive processes the female has a higher mineral requirement than the male.

3. Worms

No studies related to the determination of specific mineral requirements of worms appear to have been made, and as is so often the case, it is possible only to speculate on probable trace element requirements by comparing similar enzyme systems in other species. Unfortunately, studies of proteins in annelids appear to be restricted in the main to major body constituents. Annelids have, in common with animals of other phyla, an arginase which may be presumed to be Mn-activated. Co or Mn activate annelid phosphokinases. The occurrence of cytochrome *c*, and a cytochrome oxidase has been reported for a number of species [236] suggesting a requirement for Cu. Substantiation of a Cu requirement is indicated by the occurrence of ceruloplasmin in annelid blood [69]. The occurrence of a copper-containing luciferase in the bioluminescent earthworm *Diplocardia longa* has been reported [21]. Uricase, however, appears to be absent from annelida. Oxidation of purines to uric acid occurs in this phylum and, although there is not conclusive evidence for the presence of xanthine oxidase in each species examined, it seems reasonable to conclude that a Mo requirement for this enzyme exists.

It seems most likely, therefore, that the annelida have a requirement for Cu, Mn and Mo. A requirement for Zn also appears certain in view of the general requirement for this element for phosphomonoesterase activity.

B. Vertebrates

1. Mammals

Studies of qualitative requirements of mammals for trace elements have been confined to laboratory and domestic species. Requirements for Cu, I, Mn, Se and Zn appear well established for all mammalian species studied. The requirement for Co (as vitamin B_{12}) can, in ruminant animals, be satisfied by the metal itself because rumen bacteria carry out the necessary synthesis of vitamin B_{12}. A requirement for Mo is inferred from the occurrence of the Mo-activated enzymes, xanthine oxidase and sulphite oxidase. With the exception of a study on lambs [83] no uncomplicated Mo deficiency has been produced in any mammalian species even when highly purified diets have been used. The essentiality of F in mammals appears generally to be related only to the well-known reports of its effect in reducing dental caries, although a study by MESSER *et al.* [211] of the effects of a low fluoride intake in mice showed that an impairment in the fertility of the females occurred over two generations. Since his demonstration of the essentiality of Se, SCHWARTZ and co-workers [302, 303, 309] have shown that Si, Sn and V in the diet of rats maintained under ultra-clean environmental conditions all possess growth-promoting properties. No specific requirement for Ni has yet been demonstrated in mammalian species. A requirement for Cr is now established beyond all doubt [207, 308]. The initial demonstration by MERTZ and ROGINSKI [208] that Cr acted as a potentiating factor in insulin function and that Cr administration can effect the normalisation of impaired glucose tolerance in some animals led to the development of studies devoted to the elucidation of the character of the *'glucose tolerance factor'*.

2. Birds

As is the case with mammals, nearly all studies on mineral requirements for birds have been done in domestic species (of fowl or duck). Qualitative differences in trace element requirements between mammals and birds do not appear to exist, although, quantitatively, birds appear to have a much higher requirement for Mn than do mammals. The effects of deficiencies of Cu, Mn and Zn in domestic poultry have been investigated on numerous occasions. A growth response to supplementary Mo in a low Mo diet was observed in chicks and turkey poults [270], and extra Mo was needed in the diet of chicks which had been supplemented with W [134, 175]. Se is essential in the diet of poultry as in mammals. The essentiality of V and probable necessity for Ni to chicks has been demonstrated by the studies of HOPKINS and MOHR [138] and NIELSEN [237], respectively. No investigations appear to have been carried out on the essentiality or otherwise of Cr for birds. A requirement by the chick for Si was demonstrated by CARLISLE [42]. Birds require Co as cyanocobalamin, as do non-ruminants.

3. Fish

No specific studies on trace element requirements of fish appear to have been carried out and, in this regard, there exist considerable difficulties in working with animals living in a wholly aquatic environment. In addition, the development of suitable test diets adequate in other nutrients is not as advanced as for avian and mammalian species.

Speculation as to requirements for Cu, Zn, Mn and Fe would be warranted by the occurrence in fish of enzyme systems dependent on these metals, but only in a few instances have positive identification of metal-dependent enzymes been made. Recent studies indicate a dependence on Mn for a dipeptidase in cod muscle [234] and arginase in salmon liver [61].

The occurrence of Cu, Zn, Mn and Fe in the ovaries of tuna and other species was investigated by ESTABLIER [87] who showed that, in tuna, different levels of Zn in these tissues occurred according to whether or not the fish had spawned. Additions of Cu, Co, Zn and Mn to the water in which white Amur roe were being incubated showed that increased yields of normal embryos were obtained by addition of traces of Zn during the early stages of incubation [285]. In sturgeon fry, a study of the dynamics of Cu, Zn and Mn during ontogenesis showed that Cu content reached a peak at about 22 days old and Zn content a maximum at 42 days, Mn content was greatest at hatching [252].

4. Amphibia and Reptiles

In common with fish, amphibia and reptiles have not been studied as far as trace element requirements are concerned. In contrast, a considerable effort has been made in elucidating metabolic pathways in amphibia, no doubt as a result of the challenge posed by the metamorphic process in this class. Possible changes in the utilisation of nutrients at differing stages of metamorphosis would make difficult assessments of nutrient requirements, not least in regard to trace element requirements. The occurrence of such enzymes as uricase, monoamine oxidase, arginase, xanthine oxidase and glutamic dehydrogenase in various species of *Rana* and in *Xenopus* [15] suggest that amphibia have requirements for Cu, Mn, Mo and Zn. Speculation as to further requirements, e.g. for Se, Cr, F, and Sn, would be unrewarding. Requirements for vitamin B_{12} and thyroid hormone function indicate positive requirements for Co and I.

It appears reasonable to assume that mineral requirements of reptiles may be similar to those of avian species.

III. Functional Requirements

A. Differentiation and Pre-Natal Development

Bearing in mind the frequency with which trace element deficiency is invoked as a possible cause of poor reproductive performance in animals of agricultural importance, the dearth of information on trace element requirements for early development is surprising. Admittedly, this is a difficult field of study where the problem of being able to clearly differentiate between the role of possible maternal stores of the element and the more direct impact of changes in dietary intake. Adding some complexity to this situation are recent observations that major changes may occur in the way in which a dietary trace element may be utilised by the dam as pregnancy proceeds.

Minimum requirements for Cu for satisfactory gestation or production in mice, rabbits, pigs, poultry, goats and man are unknown. The results of a recent study by WILLIAMS and MILLS [not published] suggested that approximately 3 mg Cu/kg diet was adequate for normal gestation in the rat. In sheep adequate pre-natal development, as judged by the absence of ataxia in lamps, may be secured by 4–6 mg Cu/kg dry matter in pasture when antagonism by other elements is absent or minimal [351]. The minimum requirements for cattle, under optimum conditions, is probably about 4 mg Cu/kg dry matter.

It was shown by EVERSON *et al.* [89] that the guinea-pig required less than 1 mg Cu/kg of diet for development of the young, and reproduction appeared normal, although later development of the young was adversely affected.

Assessments of Mn requirement for satisfactory gestation or production appear to be available only for pigs, poultry and cattle. GRUMMER *et al.* [120] found a slightly improved reproductive performance when pigs were fed on a diet containing more than 12 mg Mn/kg. More than 10 mg Mn/kg was found by BENTLEY and PHILLIPS [24] to be required by dairy cattle to prevent leg defects in new-born calves, and ROJAS *et al.* [274] concluded that, for maximum fertility, the diet of cows should contain more than 16 mg Mn/kg. Requirements of poultry for Mn for normal egg production and hatchability of eggs appears fairly well established at 40 mg Mn/kg, although ATKINSON *et al.* [13] suggested that turkey hens may require between 54 and 108 mg Mn/kg.

The effect of differing Zn supplies on gestation and post-natal survival of the rat was investigated by WILLIAMS *et al.* [367]. It was concluded that 12 mg Zn/kg diet was necessary for satisfactory maintenace of pregnancy, in a semi-synthetic diet free of known Zn antagonists. Using breeding sows, HOEKSTRA *et al.* [136] found an improvement in litter size when additional Zn was added to corn-soyabean meal rations containing 30–34 mg Zn/kg. POND and JONES [258] found no improvement from additional Zn in the performance of pigs fed on a similar diet containing 35 mg Zn/kg. The presence of high levels of Ca in these diets (1.6 and 1.4%, respectively) precludes any precise estimate of basic Zn requirements for gestation in the pig. Basic requirements for Zn for normal pregnancy in other species are unknown.

B. Growth and Post-Natal Development

The effects of *in utero* malnourishment, whether of trace elements or other nutrients, cannot be disregarded when considering the later development of the animal. The degree to which an *in utero* deficiency affects the young animal depends on the species, inasmuch as birth occurs at differing stages of maturation in different species. As a consequence the development of various organs, especially the brain, is more or less affected by *in utero* malnutrition according to the stage of development at which birth takes place. In all species, however, the rate of post-natal growth in terms of proportionate increase in weight is at a maximum after birth and it would be expected that susceptibility to nutrient deficiency would be greatest at this time. This appears to be borne out by the studies so far carried out in trace element work, and in field observations.

Cu requirements for growth of rats [223], and pigs [350], have been studied. The requirement for the rat is satisfied by a level of 3 mg Cu/kg diet and for pigs appears to be in the region of 5 mg Cu/kg diet, although no precise estimate can be given for these or other species, as other dietary factors – such as type of protein source, calcium intake or the presence, in the feed of ruminants, of excessive amounts of Mo or sulphate – may increase the demand. No critical experiments on determining Cu requirements of poultry, sheep or cattle appear to have been done, but such information as is available suggests that, in the absence of adverse dietary factors, the requirement for growth in most species can be met by a dietary Cu level of about 5 mg/kg. Calves, however, may require as much as 10 mg/kg of diet.

In many species, man included, neonatal liver Cu concentrations are higher than in the mature animal, the exceptions being sheep and cattle. In sheep, liver Cu concentrations rise with age, whereas in cattle only a relatively small change in concentration occurs throughout life. The consequence of high Cu levels in the liver of the new-born is that no effect on growth becomes apparent when animals are fed on Cu-deficient diets until such hepatic stores are exhausted. This is an important consideration when assessing the availability of Cu in feedingstuffs. Animals which are born to dams in which Cu reserves have been depleted will not possess high neonatal stores and early growth and development will be swiftly affected.

In contrast, metabolic stores of Zn are small. Irrespective of the Zn status of the mother, a continual intake of Zn is necessary for post-natal growth and development. Growth of the young animal ceases shortly after the withdrawal of Zn from the diet; in the rat this occurs within 4 or 5 days, and is followed by a period in which the animal maintains a constant body weight [368]. Repletion of the diet with Zn results in a rapid resumption of growth. The inclusion of graded levels of Zn in the diet of young rats resulted in a near-linear relationship between the rate of growth and dietary Zn concentration up to a level of 12 mg Zn/kg diet. Experiments with lambs and calves [224] indicated that about 7 and 8 mg Zn, respectively, per kg diet were required to maintain a good rate of growth, although these levels were not sufficient to maintain plasma Zn concentration at the normal level. Normal growth of pigs was obtained on a casein-based diet containing 10 mg Zn/kg by MILLAR *et al.* [214]. It must be emphasised that these results were obtained by the use of purified diets and are *not* directly applicable to natural feedingstuffs which may contain high levels of Ca and phytic acid which, by their effect on rendering Zn less available, may increase Zn requirements by a factor of three or four. This effect appears to be of less consequence in birds and appears not to apply to adult ruminants. Domestic poultry have a higher requirement than do mammals and this is in the region of 30–40 mg Zn/kg diet.

Birds also appear to have a high requirement for Mn in comparison with mammalian species, which has been attributed to a relatively poor absorption of Mn from the avian gut. The lack of any precise figures for the growth requirement for Mn in most species prevents the making of any valid comparisons. Diets low in Mn (1 mg Mn/kg or less) have not allowed normal growth to occur in mice, rats or rabbits [317, 326, 354], whereas a normal growth rate was observed in the case of pigs [158, 256, 257] although Mn depletion of tissues occurred. It is not clear whether these differences reflect any alteration in availability of Mn stores in the different species.

The requirement for cobalt in ruminants (and here sheep and cattle may be considered together) depends on the results of field studies in Australia and New Zealand which indicated that a pasture level of 0.07 or 0.08 μg Co/kg dry diet was probably adequate. As pointed out by UNDERWOOD [351], more precise estimates of minimal Co requirement are difficult in view of the influence of other variables such as grazing habits and contamination of feed by soil.

Iodine requirements by different species have been the subject of much study and provide a clear example of the frequent need to consider the nature of other dietary components when attempting to assess requirements for a trace element. Assessment of I requirements in man are cited by UNDERWOOD [351] as in the range of 44–200 μg/day; that for the rat [182] at 1–2 μg/day (at which level no enlargement of the thyroid was observed), and dietary requirements for growing poultry have been given as near 1 mg/kg diet [351] or 0.03 mg/kg for normal growth or 0.3 mg/kg for normal thyroid structure [110]. The effects of nutritional variables on the requirement for I and on the relationships which exist between I levels, incidence of goitre and the occurrence of other potential interfering substances such as thiocyanates, arsenic, fluoride or Co and Mn have been succinctly summarised by UNDERWOOD [351].

Se deficiency has been associated with 'ill-thrift' and poor growth in lambs, retarded growth in chicks and in Japanese quail. 'Ill-thrift' syndrome, responsive to Se therapy, may also occur in mature sheep and cows. The quantitative requirements for Se for growth in sheep and cattle are a matter for some dispute. Se analyses in herbage cannot easily be correlated with the incidence of Se-responsive disease, and UNDERWOOD [351] has pointed out that Se requirements may be greater when high dietary sulphate levels occur. A dietary intake of 0.1 mg Se/kg of diet is suggested as being adequate for ruminant animals [3], and is presumed sufficient to accomodate the effect of other dietary variables. In the case of poultry, about 0.1 mg/kg was found adequate for maximum growth of chicks [238]. The interrelationship between Se and vitamin E (sect. V) was found by SCOTT *et al.* [312] to be important in considering the Se requirement of turkey poults. These authors found that the dietary Se requirement of these animals increased from 0.18 to 0.28 mg/kg when vitamin E was omitted from the diet.

Requirements for Si in the growing chick and rat were demonstrated by CARLISLE [42] and SCHWARTZ [303], respectively. In the first instance the addition of 100 mg Si/kg to a low Si basal diet, stimulated chick growth by between 30 and 50%. A higher requirement was found for the rat when significantly improved growth rates were not observed if less than 500 mg Si/kg diet were supplied. Requirements for V and Sn for growth of the rat were established to be about 0.1 and 1.0 mg/kg diet, respectively [302, 309].

The form in which the trace element is supplied frequently determines its biological activity. No quantitative assessment of Cr requirement can yet be established until the biologically available forms have been identified. In the case of Cr, considerable differences in availability, metabolism and function have been noted for the *tri-* or *hexa*valent inorganic salts and the unidentified organic form [208].

A growth response by lambs to supplementary Mo [83] was attributed to the effect of this element in stimulating cellulose degradation by rumen microorganisms.

C. Reproduction (and Lactation)

The maintenance of pregnancy appears, in the rat, to be dependent on the dietary supply of not less than 3 mg Cu/kg diet [WILLIAMS and MILLS, to be published]. At lower levels of Cu, oestrus cycles are not abolished. In contrast, Cu deficiency in dairy cattle is frequently associated with delayed or depressed oestrus. The minimum requirements of Cu for reproduction and lactation in nearly all species appears not to have been investigated, and it is tacitly assumed that requirements for reproduction are not dissimilar to those for growth.

For Zn, it was shown [367] that the requirements of the rat for maintenance of pregnancy and post-natal survival of the offspring were met by a concentration of 12 mg Zn / kg diet, when a purified diet was used, this concentration being adequate for normal growth rates. Adequate dietary Zn is required for the maintenance of oestrus cycles in the rat [145]. Minimal requirements for reproduction and lactation in other mammalian species appear not to have been investigated. One study by SWENERTON and HURLEY [345] suggests that, in male rats, the requirement for Zn for correct testicular development and spermatogenesis is considerably greater than that required for growth, and these authors suggested that more than 60 mg Zn / kg diet was probably necessary for complete testicular function. This figure is based on the results of experiments in which an isolated soy protein was used and which may have decreased the availability of the dietary Zn. More than 17 mg Zn / kg diet was necessary for correct testicular function in rams [353].

There is ample evidence that, in both male and female, defects in reproductive function may occur when dietary Mn levels are not sufficiently depressed to adversely affect growth. Minimum dietary Mn requirements for reproduction in rats, rabbits, mice, pigs, sheep, goats and cattle are unknown. For most species, reproductive defects are not observed when rations contain 40 mg Mn / kg. In the case of cattle, however, HARTMANS [126] has shown that no improvement in fertility, growth or other function was obtained by supplementing with extra Mn, cattle which were grazing pastures containing 25–30 mg Mn / kg.

Poultry may require as much as 50 mg Mn / kg for optimum reproductive performance, but one report [13] suggests that turkey hens may require between 54 and 108 mg Mn / kg. As with other elements, requirements are influenced by other dietary factors. High Ca intakes markedly decrease the availability of both Zn and Mn. In this regard the generally high rates of inclusion of Ca in rations of laying birds substantially increases the amounts of dietary Zn and Mn required to obtain optimum performance in terms of both growth and reproductive capacity.

Assessments of *minimal* requirements for adequate reproduction have not been made for other trace elements and, in reproductive studies involving mineral requirements, the male animal appears to have been particularly neglected.

IV. Effects of Deficiency and Toxicity

Deficiencies of *cobalt* have been reported only in ruminant animals. Weakness, anaemia and body wastage (enzootic marasmus) in sheep and cattle grazing certain areas of Western Australia were shown to be reversible by oral Co therapy [196, 352]. The disease state, in all its manifestations, is clearly accounted for by the depletion of vitamin B_{12} stores and is, therefore, outside the scope of this text. In addition, however, there appears to be in the sheep (and, to some extent, cattle) a specific requirement for Co in preventing the appearance of neurological dysfunction in those animals exposed to the toxic effects of ingesting the grass, *Phalaris tuberosa* [351]. Degenerative changes in the nervous system are prevented by Co administration, whereas vitamin B_{12} is ineffective. Neither Co nor vitamin B_{12} is effective against the acute form of the disease. That regular dosing of sheep with Co prevented the appearance of 'Phalaris staggers' was demonstrated by LEE and KUCHEL [179], who first observed the relation between the disease and incipient Co deficiency. The mechanism by which Co protects against *Phalaris* toxicity is unknown.

Cr deficiency has been identified as responsible for reductions in growth rate of rats and has adverse effects on the metabolism of glucose [307, 308], lipids [291, 292, 300] and protein [271, 273]. Impaired glucose tolerance in rats was corrected by administration of compounds of tri-valent Cr, whereas hexavalent Cr was inactive. Rats fed a low Cr diet developed lesions in the eye [272] manifested as a sequential loss of brilliance, dilatation of the blood vessels and the development of corneal opacity. Rats fed on diets containing white sugar (low in Cr) had elevated blood cholesterol levels compared with those receiving brown sugar or white sugar plus Cr [291]. Protein synthesis in rat heart was adversely affected by Cr deficiency.

The studies of ROGINSKI and MERTZ [271–273] showed that amino acid incorporation into protein in Cr-deficient rats was stimulated by tri-valent Cr, although the effect was not a general one on all amino acids.

Longevity of male rats and mice was increased in the presence of dietary Cr. This effect was not observed in female rats or mice.

In human studies the possibility of Cr deficiency occurring in old people and during pregnancy is indicated by the studies of DOISY *et al.* [78], of TIPTON [348] and HAMBIDGE [122]. Cr concentrations are higher in foetal and neonatal tissues and generally decline with age. Repeated pregnancies may result in depletion of maternal Cr reserves with adverse effects on mother and later offspring. As pointed out by HAMBIDGE, it is necessary to determine whether the decline in Cr status in early life is a normal physiological process; the possibility exists that it may be indicative of depletion below an optimal Cr nutritive status.

Cu deficiency has been associated with a wide variety of defects in many species. These include foetal death and resorption when the deficiency is severe [81], anaemia, depigmentation of hair and wool and, in the latter, a loss of crimp. In addition, disorders of bone (rarefaction, brittleness) and of the gastrointestinal tract occur. Cardiovascular defects and heart failure are also common effects of Cu deficiency. Demyelination of the brain occurs in severely Cu-depleted lambs and neonatal ataxia has been reported in sheep, goats and guinea-pigs.

The anaemia of Cu deficiency appears to result from a defect in erythrocyte production and maturation. Species-specific differences occur in the character of the morphologic changes which develop in the Cu-deficient erythrocyte; a normocytic and normochromic anaemia appears to occur only in chicks [199] and dogs [191]. In other species (rats, rabbits, pigs) and in young lambs the anaemia is microcytic and hypochromic [23, 171, 325, 327], whereas in adult sheep it is macrocytic [197], and in cattle it is macrocytic and hypochromic [23, 60]. Defects in Fe absorption and transport have been related to a decline in serum of the activity of the Cu-containing enzyme, ceruloplasmin (ferroxidase) in Cu deficiency. The hypothesis that Fe transfer and hence availability is dependent on the oxidation by this enzyme of ferrous to ferric iron is compatible with the findings of LEE *et al.* [178] that in Cu-deficient pigs no defects were found in haeme biosynthetic pathways, and these authors suggested that Cu is essential either for the normal metabolism of iron or for synthesis of globin. Defects in protein synthesis do not, however, appear to occur in the Cu deficiency state [351].

Depigmentation of hair and wool has been observed in Cu-deficient rats, cats, dogs, rabbits and ruminants. The occurrence of a hair defect (Menkes' kinky hair syndrome) in humans is attributed to a defect in Cu absorption; the duodenal mucosa contained abnormally high levels of Cu [63].

A defect in elastin biosynthesis in Cu deficiency appears to be associated with reductions in the tensile strength of the aorta. Histological examination of the aortas of Cu-deficient chicks [239] showed a derangement of the elastic tissue. Death in these animals resulted from ruptures of major blood vessels. Similar findings were made in pigs [43] and guinea-pigs [89].

The involvement of Cu in collagen formation, in a manner similar to that of elastin cross-linking, is associated with marked differences in the character of this protein extracted from the bones of Cu-deficient chicks [282]. An increase in the solubility of collagen from Cu-deficient bones, indicative of alterations in structural integrity of the protein, suggests that a major defect in bone formation in Cu deficiency is a failure in the structural stability of the organic matrix.

Neonatal ataxia in lambs, goats [129] and guinea-pigs [89] has been reported; that in lambs has been closely studied. Since the original characterisation of the disease by INNES and SHEARER [156] it has been shown that in the brain stem and spinal cord marked degenerative changes occurred in the large nerve cells [16, 17]. FELL *et al.* [93] and MILLS and FELL [225] demonstrated degeneration of the nuclei of motor neurones in the red nucleus, and showed that cytochrome oxidase activity was significantly depleted in the brain of sway-back lambs and, in addition, that a marked fall in brain Cu levels to less than 3 μg Cu/g dry weight occurred in these animals [227]. Histochemical confirmation of the low cytochrome oxidase activity in brain tissue was demonstrated by FELL *et al.* [92].

In human studies, Cu deficiency symptoms may arise as a result of metabolic disturbances associated with generalised malnutrition. In 'protein-calorie' malnutrition, there is evidence that mineral deficiency states may be general; the liver of a child dead from this cause was found to contain significantly less Cu, Zn and Mn than normal [180]. The Cu concentration in abdominal wall muscle and Mn concentration in heart muscle was significantly depleted. The effect of kwashiorkor in adversely modifying the concentrations of Cu, Fe and Fe-binding capacity in the serum was studied by LAHEY *et al.* [170]. A similar finding was made by ABDEL-SALAM *et al.* [1]. Low serum Cu values were determined in infants born to malnourished mothers and, in addition, a significant correlation was found between birth weight and serum ceruloplasmin activity [169]. A reduction in ulnar vein blood Cu was observed in women showing uterine inertia during childbirth [163].

A Cu-responsive neutropenia was observed in Peruvian infants who showed, in addition, 'scurvy-like' bone changes and depressed serum Cu levels [52, 53]. In more than one third of infants and children exhibiting severe malnutrition and chronic diarrhoea a diagnosis of Cu deficiency was made by GRAHAM and CORDANO [115]. That Cu deficiency in children may occur even in developed countries is indicated by the studies of STURGEON and BRUBAKER [337] and SCHUBERT and LAHEY [301] in two major cities of the United States. A study of prematurely born infants by SULTANOVA [340] showed that there is marked depletion of Cu reserves in both liver and spleen. In view of these studies there appears to be a strong case for further detailed examination of the Cu status in early post-natal life. A comprehensive review of Cu deficiency in man has been made by CARTWRIGHT and WINTROBE [44].

Congenital ataxia is a characteristic feature shared by the young of all species in which a deficiency of *Mn* has been imposed upon the maternal organism. In studies on the rat HURLEY *et al.* [142] demonstrated that the requirement for Mn became critical between 15 and 16 days of pregnancy if ataxia in the new-born were to be prevented. It was shown that a defect in vestibular function and in development of the otoliths, associated with the ataxia, occurred also in guinea-pigs [318] and mice [86]. The discovery by HURLEY and co-workers [84, 85] that, in a strain of mice (pallid, *pa* mutant) supplementation of the diet by extra Mn during pregnancy prevented the defective development of the otoliths which results, in this strain, from the presence of the mutant gene, appears to be the only known instance of an agent involved in both phenocopy induction and prevention. The further discovery that otolith development in mice in response to Mn deficiency is genetically linked [141] is of considerable importance, and suggests that Mn may be associated with the transmission of hereditarily linked characteristics. In this connexion ANKE *et al.* [6] have shown in cattle and goats subjected to Mn deficiency, significantly more male than female offspring were produced and that it became possible to select animals which were capable of utilising Mn more efficiently than was initially the case. The same workers [132] also showed that the birth weight of Mn-deficient kids was lower than that of controls and in this experiment there emerged families less susceptible to the deficiency of Mn.

Skeletal deformities resulting from Mn deficiency have been characterised in many species. In the chick, perosis [364] is characterised by the swelling of the tibio-metatarsal joint and slipping of the gastrocnemius tendon. In mammalian species the bone deformities may be apparent as bowing or shortening of the limbs, lameness, difficulties in walking or joint pains. Although species-specific differences in response are notable, UNDERWOOD [351] has pointed out that these may also depend upon age of animal and duration and severity of the deficiency state. As suggested by this author a defect in calcification can be ruled out as a causal factor in the Mn deficiency induced abnormalities but rather that the lesions arise as a result of the reduction in chondroitin sulphate content and hence rigidity of the organic matrix of cartilage and connective tissue. In poultry, eggshell thickness is decreased in Mn deficiency and it has been shown by HILL and co-workers [135, 188] that this defect occurs when Mn-deficient diets are fed up to point-of-lay but not afterwards. They concluded that Mn was needed for development of physiological functions necessary for complete shell formation. As in cartilage formation, Mn deficiency depressed hexosamine and uronic acid content of the organic matrix of the shell, although uronic acid content rose after three months of laying on a low Mn diet [187].

Impairment of reproductive function in both male and female occurs as a consequence of Mn deficiency. In the many years since initial demonstrations [161, 242, 360] that Mn deficiency is associated with reproductive disorders and perinatal mortality no firm evidence has appeared as to the identification of the mechanisms involved. In the female mammal, Mn deficiency is associated with depressed oestrus and impairment of development of the young. Cows [24, 274], goats [5] and pigs [120] have all been shown to require an adequate supply of Mn for optimal sexual functioning, and decreases in litter size have been observed in pigs [256, 257] and guinea-pigs [88] as a result of Mn deficiency.

In male rats and rabbits, Mn deficiency is associated with a lack of libido, degeneration of seminal tubules and lack of spermatozoa [30, 326]. Libido, mating and fertility were not affected in Mn-deficient goats, although motility of spermatozoa and ejaculate density were decreased and a high incidence of abnormal spermatozoa were found [119]. These apparent species-specific differences may again reflect the severity and duration of the deficiency state.

Iodine deficiency in animals is essentially linked with its effect on thyroid hormone function and is therefore mainly outside the scope of this text. Interactions between I and other elements have been reported; e.g. arsenic has been shown to induce goitre in rats [316] and both Co and Mn appear necessary for thyroid hormone synthesis [27]. Excess Mo administered to rats slowed growth, inhibited spermatogenesis and caused hyperaemia of the liver [228]. These toxic effects of Mo were abolished by the simultaneous addition of I, but the effects of Mo and I together were to cause hyperplasia and dystrophic changes in the thyroid.

Reduced activity of cytochrome oxidase was demonstrated in leucocytes of rats and guinea-pigs fed a diet deficient in I [375]. In addition the leucocytes contained less glycogen, peroxidase activity was low and their phagocytic and bactericidal properties were adversely affected. In human patients suffering from goitre the concentration of Mn in the thyroid was higher than in euthyroid subjects [289] and Mn levels in foodstuffs in areas where goitre was prevalent were higher than in other regions. Interactions between I, Co and Mn appear to merit close investigation, as do those of the effects of Ca and F on I metabolism where a considerable amount of contradictory data exists.

Fluorine deficiency studies, based on specially designed low F diets, have in the main failed to produce evidence of a specific requirement for F. Contradictory evidence regarding the role of F in the prevention of dental caries in rats is summarised by UNDERWOOD [351]. The effects of F on dental caries and bone osteoporosis have more recently been investigated; SRIRANGEREDDY and RAO [334] were able to show that 135 mg F/kg diet did, after a considerable period, mitigate to some extent against the development of low Ca-induced osteoporosis in rats, but 45 mg F/kg diet did not. Dietary F did not prevent osteoporosis in Ca-deficient beagle dogs [133]. In another study 10 mg F/kg diet reduced cavitation of the fissures of rat molars [190]. The degree of fluorosis and incorporation of F into bones and teeth of mice under conditions of high F intake was shown to depend on F concentration in the water rather than on daily F intake [283]. In a study of over 20,000 Japanese children, IMAI [153] showed that dental caries was lowest in regions where the F concentration in drinking water lay between 0.3 and 0.39 mg/litre. In particular, the incidence of dental caries on one Japanese island fell from 67 to 58.2% as water F concentration rose from 0.0–0.099 to 0.300–0.399 mg/litre [155]. In a third study, IMAI and KITAMURU [154] showed that dental caries in children was higher when the water contained from 1.0 to 8.6 mg F/litre than when the F concentration was from 0.1 to 0.8 mg/litre. It would appear, therefore, that effects of F on the hard tissues of the body depend not only on concentration in food or water, but on the mode of intake and other dietary factors. Mg deficiency in the guinea-pig, which leads to calcification of the kidney and mortality, was found to be alleviated by the presence of 100 mg F/kg diet [240].

The effects of low F diets on reproduction in mice were reported by MESSER *et al.* [210, 211], and this work provides the first evidence for the essentiality of F in processes other than those concerned with calcified tissues. A progressive decline in the fertility of female mice was noted in succeeding generations maintained on a low F diet (0.1–0.3 mg F/kg); although growth rate and litter size were not influenced by F deficiency, the proportion of animals producing litters declined. Fertility was restored by the addition of F to the deficient diet.

Selenium deficiency, originally associated with hepatic degeneration in rats and exudative diathesis in chicks [305, 306, 310] has since been shown to be one of the dietary factors responsible for the occurrence of 'white muscle disease' in young ruminants and foals [124, 203, 230, 266], a reduction in growth rate and reduced fertility in rats and sheep [125, 202, 203, 333] and the occurrence of hepatosis dietetica in pigs [82]. Sperm motility was markedly decreased in Se-deficient rats [202, 373]. Squirrel monkeys lost weight and developed alopecia, myopathy and hepatolenticular degeneration [231]. Alopecia was observed in the offspring of female rats deficient in Se [146, 202] and skeletal and smooth muscle necrosis occurred in Se-deficient ducklings [374].

White muscle disease, characterised by the development of white streaky areas in striated muscle (particularly in those of thigh and shoulder muscles) is essentially a disease of the young growing animal, and assumes particular economic importance in lambs and calves. Necrosis of muscles has been observed in nearly all species affected by Se-deficiency, although in chicks and young pigs the simultaneous occurrence of exudative diathesis in the former and hepatic degeneration in the latter tend to obscure the effects on muscle integrity [351].

Suboptimal Se intakes in grazing ruminants range from relatively severe deficiencies as may occur in New Zealand [351] to marginal deficiencies under Scottish conditions, as observed by BLAXTER [26]. Intermediate deficiencies of Se are presumably responsible, to some extent, for the loosely characterised 'ill-thrift' symptoms of sheep and cattle, although in these instances there is often an association with Co-deficiency and parasitic infections. That Se may be involved in the animal's ability to overcome the effects of infection is suggested by the work of SPALLHOLZ *et al.* [329] who showed that mice given Se at levels above those regarded as nutritionally adequate had an improved immune response. Se may also be involved in processes concerned with absorption or the subsequent fate of other minerals, as THOMSON and LAWSON [347] showed that ewes and lambs grazing low-Cu pastures had improved blood and liver Cu status after Se administration.

Species differences appear to exist in the response of animals to the effects of Se deficiency on fertility. While both male and female rats became infertile over successive generations when exposed to Se deficiency, in sheep defects in oestrus or ovulation were not observed, but embryonic mortality became high [351]. Testicular defects were not observed in Se-deficient ram lambs [37].

Although adverse effects on growth rate are frequently associated with Se deficiency, the occurrence of severe liver lesions in young pigs (hepatosis dietetica) may occur without any serious effects on growth to the time when sudden death may occur.

Growth rate was affected in young chicks by Se deficiency [238]. Initial deficiency symptoms in day-old chicks exposed to Se deficiency occurred in six days [118] with degeneration of the exocrine pancreas. The typical features of exudative diathesis which include the accumulation of fluid throughout the body, appear initially as an oedema on the breast, neck and wing. Later, subcutaneous haemorrhages appear which may account for the fluid accumulation. Mortality is frequently high. The addition of antioxidants or vitamin E to the diet reduced both the incidence and severity of the disease, but only Se was completely effective [248, 305]. The hatchability of eggs and viability of the young chick was also adversely affected by Se deficiency [157, 284].

Some integration of the apparently disparate observations made upon the effects of Se deficiency in animals is indicated by the studies of SPRINKER *et al.* [333] who observed, in rats, lesions similar to those of Se-responsive exudative diathesis in chicks, unthriftiness in lambs and calves, and infertility in ewes; it was concluded that the primary effect of Se deficiency is abnormal vascularisation associated with endothelial degeneration.

Zinc deficiency in all species studied results in growth failure, impaired reproductive function, defects in skin and modified hair, wool or feather growth. Bone lesions may occur, particularly in birds.

In rats, a cessation of growth occurred within 5 days of feeding on a Zn-deficient diet [368] followed by a 5- to 6-week period during which the animals maintained a constant weight. This was accompanied by a regular variation in food intake, alternating in cycles of high and low intake with a periodicity of about 3½ days [49, 368]. Plasma Zn levels declined rapidly during the early stage of Zn deficiency in rats, sheep and calves [79, 80, 224], but there was no evidence to suggest that this change was associated immediately with effects on food intake or growth rate. Subsequent changes in the physical appearance of the animal involved loss of hair [18, 80, 345] or wool [224, 353] and the development of parakeratotic lesions of the skin, tongue and oesophagus [18, 97, 349].

Effects of Zn deficiency on reproductive function involve almost every phase of the reproductive cycle in both male and female. Testicular atrophy, hypogonadism and failure of spermatogenesis or production of malformed sperm has been reported in rats, goats, sheep, cattle and man [19, 189, 213, 216, 218, 235, 243, 263, 264, 353]. Female rats made severely Zn-deficient became anoestric [145] and when Zn deficiency was imposed during pregnacy or shortly after mating, difficulties in parturition were observed in rats and rabbits [7–9] and the animals produced malformed young [145]. In rats, calves and man [217, 259, 260, 287] a subnormal Zn status was associated with retarded healing of wounds. The occurrence of dermatitic lesions in the Zn-deficient animal may, in some instances, be the result of minor abrasions which fail to heal as rapidly as in a normal animal. Defects in keratogenesis in the sheep were indicated by structural changes in new horn growths [224, 353] while in breeds normally hornless, development of 'buds' in the position of horns was reported [224]. In chicks, turkey poults, pheasants and quail [168, 241, 311, 331] poor feathering occurs. Integumental and skeletal disturbances were observed in chicks hatched from eggs of Zn-deficient hens [166]. Leg deformities were observed in Zn-deficient calves [215]. Effects of transitory Zn deficiency during pregnancy in the rat have been investigated [143, 144, 356, 357]. Congenital abnormalities and decreased post-natal survival occurred, the brain and central nervous system being much affected. Adverse effects of Zn deficiency on the behaviour of the offspring of mildly Zn-deficient rats was reported [41].

Inhibition of development of a variety of animal tumours by dietary Zn deficiency appears to be well confirmed [20, 71, 204], although in contrast, supplementary dietary Zn was shown to inhibit the development of tumours induced by DMBA (9,10-dimethyl-1,2-benzanthracene) in the hamster cheek pouch [261].

Toxicity studies on trace elements, as applied to animal nutrition, will be discussed only briefly, inasmuch as concern need only be felt for those instances where the elements are ingested along with foodstuffs.

Susceptibility of species to ingestion of toxic quantities of the trace elements varies. Chronic *Cu* poisoning may arise as a result of contamination of feeds, or by excessive absorption (in the case of sheep) of Cu from natural pastures. The effect particularly on sheep, of grazing high Cu pastures is to cause the storage of very high levels of Cu in the liver which then may be suddenly released to induce a 'haemolytic crisis'. Rapid kidney damage occurs and high blood urea values are observed. Chronic Cu poisoning in sheep has been recently reinvestigated by GOPINATH *et al.* [113]. Haemolysis, and the associated jaundice, is not a normal characteristic of Cu poisoning in other species.

Cows appear to be more tolerant than sheep to a high Cu intake. Up to 2 g copper sulphate given daily to adult cows produced no adverse symptoms [95], although 5 g daily was fatal to a steer [165]. Differences in the apparent susceptibility of calves to Cu poisoning [95, 315] may well reflect variations in age or experimental treatment, and also the presence in the diet of other factors, such as Fe level, Zn, Mo and inorganic sulphate concentrations. As pointed out by UNDERWOOD [351], intakes of Cu alone can be misleading as indices of Cu toxicity and Cu deficiency. A study by SUTTLE and MILLS [343, 344] showed that increasing dietary levels of Fe and Zn in pig diets containing as much as 450 mg Cu/kg completely protected against Cu poisoning. In other mammalian species and birds, Cu appears to have a fairly low order of toxicity. Rats remained healthy on diets containing

500 mg Cu / kg [29]. Ducklings accumulated much more Cu in their livers than did chicks when fed diets containing about 200 mg Cu / kg, but there was no effect on growth rate [371].

Cr toxicity is dependent on the form of the element which is administered, hexavalent Cr being more toxic than the trivalent form. This may be related to the greater absorption of Cr^{6+}. Levels of Cr greatly in excess of requirement have produced growth depression and damage to liver and kidney in experimental animals [351]. A claim that 10 mg Cr / kg diet stimulated growth in the rat has been made by CHATTERJEE *et al.* [47].

Co induces polycythaemia in many species but has a low order of toxicity. UNDERWOOD [351] cites observations of PERRY and SCHROEDER who treated hypertensive patients with 50 mg oral $CoCl_2$ per day for 10–65 days. No toxic effects were observed. Sheep can tolerate doses of 3 mg / kg body weight without adverse effects [22], but higher doses depress appetite. Cobalt, as $CoSO_4$, in concentrations of 0.05–5.0 mg / litre was found deleterious to developing White Amur roe [285].

Fluorine toxicity in all its manifestations, is fully discussed by UNDERWOOD [351]. Tolerance to fluoride depends on the form administered and species of animal; birds, rats and mice, being less susceptible than other animals, may tolerate intakes of hundreds of parts per million of dietary F. Maximum safe levels for cows, on the other hand, may be of the order of 50 mg F / kg diet.

Tolerance to intakes of *I* in excess of requirement differs according to species [4, 10, 11, 195, 251]. Up to 500 mg I / kg of diet fed throughout pregnancy to female rats had little effect on litter development. Increasing mortality of the young occurred with increasing I level in the diet. Male rats and gestating sows were not affected by levels of 2,500 mg I / kg diet, while at 250 mg I / kg, higher mortality of the young was observed in rabbits. Laying hens appeared susceptible and adverse effects on egg production was observed at intakes of 625 mg I / kg and an increase in embryonic mortality at dietary I concentrations of 312 mg / kg diet. Supplementation of the diet of pregnant ewes with I at the rate of 0.05 g potassium iodide per day had a slightly deleterious effect on weight gain and reproductive performance, but no I-specific effect could be demonstrated in the thyroids [194]. Mechanisms involved in I toxicity are not known.

Interference with Fe metabolism appears to be the main effect of *Mn* toxicity. Low orders of toxicity of Mn occur in all species, growing pigs appearing the most susceptible in being affected by intakes of 500 mg Mn / kg [120]. Animals affected by Fe deficiency anaemia in early growth, e.g. lambs and baby pigs, may be susceptible to Mn toxicity at levels of only 45–50 mg Mn/kg diet [200]. These effects are probably related to effects of Mn on Fe absorption. Direct access of Mn to the body via the lungs as well as the gut is probably the cause of chronic Mn poisoning in manganese miners; in this instance it would appear that mechanisms of homeostatic control of tissue Mn levels may be overburdened by the high absorption rate, via the lung tissue [351].

Interactions between *Mo,* Cu and inorganic sulphate in ruminants and other species have been reviewed by UNDERWOOD [351]. The complexity of these interactions is still not clearly understood. Mo toxicity in all species results in depressed growth rate or falls in body weight. Diarrhoea, alopecia, sterility, anaemia, achromothrichia and bone lesions have been reported in one or more species as a result of Mo toxicity [351]. Tolerance to high Mo intakes varies enormously between species, ruminants being particularly susceptible, while pigs have shown no ill effects with concentrations of 1,000 mg Mo/kg feed. Toxicity of Mo is dependent ot a large extent on dietary Cu levels and also on the ratio of Mo to Cu in the diet. Supplementary Mo may induce a 'conditioned' Cu deficiency and, conversely, Cu has a protective action against Mo toxicity, as in non-ruminant species do high dietary contents of sulphate or S-containing amino acids.

Nickel is poorly absorbed from the gut and appears to have low toxicity. Growth retardation was observed in female but not male mice at a level of 1,100 mg Ni/kg in the diet [359]. Both sexes were affected when the concentration was increased to 1,600 mg/kg. An adverse effect of Ni on growth was observed in chicks at 700 mg Ni/kg, which was associated with adverse effects on energy metabolism and nitrogen retention [358]. Cytochrome oxidase activity was reduced in the tissues of Ni-poisoned mice and rats [359, 362], and adverse effects were also noted on the activity of malic and isocitric dehydrogenase activities. In plasma, alkaline phosphatase activity was reduced whereas ceruloplasmin was not. Increases in liver Cu and Zn concentrations were also observed [362].

The toxicity of *Se* is well established. Widespread clinical effects have been noted in all species. Chronic mild poisoning results in generalised disturbances characterised by lethargy, loss of hair, erosion of bone joints leading to stiffness in movement and lameness. Soreness and deformation of the hooves of cattle, horses and pigs have also been observed. Anaemia, liver damage and teratologic changes in embryos have been reported.

FRANKE and POTTER [103] showed that the administration of Se to experimental animals produced clinical symptoms similar to those associated with 'alkali disease' in grazing stock endemic in parts of the USA and Eire. In rats and dogs, studies on the development of Se-induced anaemia showed that haemoglobin levels may fall to 20% of normal values [229]. Decreases in egg-production, reduced fertility and the production of malformed young have been noted in many species [75, 103–105, 275, 276, 355].

The toxicity of Se is dependent on the form administered, duration of exposure and level in the diet. Minimum levels at which toxicity symptoms appear are therefore hard to define. It was suggested by UNDERWOOD [351] that minimum Se levels in the herbage of grazing stock at which toxicity symptoms may appear probably lie close to 5 mg/kg. Rats which show signs of chronic poisoning at a dietary level of 5 mg Se/kg appear to be protected by an increase in dietary protein level to 30%, which may be due to an increased production of endogenous sulphate [121]. A far more effective protection against Se toxicity is afforded by the discovery of MOXON and RHIAN [229] who showed that 5 mg/litre of arsenic in drinking water completely prevented Se toxicity in rats. A relationship, somewhat paradoxical between Se and As is indicated by the findings of MUTH *et al.* [232] that As has a preventive effect in the development of myopathy in lambs fed a Se-deficient diet.

Vanadium toxicity has been demonstrated in rats and chicks. A concentration of 10 mg V/kg diet was sufficient to reduce significantly the food consumption of growing chicks [341], while 25 mg V/kg produces toxic symptoms in rats [102].

In mice and rats given 5 mg *Sn*/litre in drinking water, a reduced longevity was observed in female rats. Fatty degeneration of the liver and minor kidney damage was observed in both sexes [293, 297].

Zinc has a remarkably low order of toxicity to mammals and birds. Ruminant animals appear more sensitive to high Zn intakes than do rats, pigs or poultry. Dietary levels of Zn between 500 and 1,000 mg/kg have been observed to depress weight gains in lambs, steers and heifers [244, 245]. Pigs, rats and poultry have been fed over 1,000 mg Zn/kg without apparent ill effect [130, 183, 250]. At higher levels, ranging from 3,000 to 5,000 mg/kg diet according to species, reduced food consumption, anaemia and internal haemorrhages have been noted. In the rat, anaemia and reductions in the activities of Cu- and Fe-containing enzymes can be overcome by the addition of extra amounts of these elements to the diet [56, 192] suggesting that the effect of Zn is an interference with absorption and utilisation of Cu and Fe. In lambs, changes in rumen metabolism were observed [246] which included a reduction in volatile fatty acid concentration and a diminution in the acetic/propionic acid ratio. The authors suggested that an effect of Zn on rumen microorganisms may explain the lower tolerance to excess Zn in ruminants.

V. Metabolism and Biological Role

Investigations on *Cr* metabolism have shown that absorption, tissue uptake and excretion are highly dependent on the chemical form of the element. Orally administered, Cr^{3+} is poorly absorbed as compared with hexavalent Cr, the latter being capable of readily passing through red blood cell membranes and becoming attached to globin. Tissue uptake of Cr is rapid, but whole-body Cr disappearance is slower [209], more being retained in bones of young rats and in spleen, testes, kidney and epididymis in older rats than in other tissues [137]. The occurrence of relatively high concentrations of Cr in the foetal or new-born animal appeared paradoxical in view of the failure to detect any significant uptake of ^{51}Cr by the foetus, following its administration to the mother. This problem has been partly resolved by the studies of MERTZ and co-workers [207–209] who have shown that an unidentified complex of Cr ('glucose tolerance factor') prepared from ^{51}Cr-labelled yeast was capable of passing the placenta [208]. Absorption studies *in vivo* by MERTZ and ROGINSKI [208] have also shown that intestinal absorption and retention, in the body of the rat, of the Cr complex obtained from yeast was many times more than occurs when Cr^{3+} is given as the chloride. The biological role of Cr is unknown, inasmuch as precise sites of action have not been located, and must await the identification of the organic Cr complex.

Cu absorption in many species is markedly influenced by other dietary components. Among those adversely affecting Cu absorption are Zn, Mo, cadmium, and in ruminant species the presence of elevated contents of sulphate or other S-containing compounds in the diet. The mechanism of Cu absorption is little understood, but a Cu-binding protein has been found by STARCHER [336] in chick intestinal mucosa. The chemical form in which Cu exists influences the degree of absorption [219–222]; water-soluble complexes of Cu extracted from herbage were more available to the rat than inorganic Cu as the sulphate. A study of the complexes of Cu and other metals in ruminant feed and intestinal contents has been made by BREMNER and KNIGHT [33, 34]. Plasma Cu is loosely bound to serum albumin and from this source reaches other tissues. Unabsorbed Cu, together with that eliminated via the bile, is excreted via the faeces – the major route of Cu elimination. An examination of the Cu-binding capacity of human alimentary secretions [111, 112] showed that bile contained a highly stable complex, and it seems that net Cu absorption in man may represent an interplay between the opposing influences of bile and endogenous low molecular weight complexes.

Of all the Cu-containing enzymes, diminution in the activities of cytochrome oxidase, ceruloplasmin and lysyl oxidase in the deficient animal have the most profound effects on its metabolism. Bone and cardiovascular defects associated with changes in the character of collagen and elastin, result from a diminution in activity of the Cu-containing enzyme, lysyl oxidase [198, 280, 281, 319] and consequent failure of cross-linking of the protein molecules.

The Cu-containing enzyme ceruloplasmin (ferroxidase I) found in serum, and synthesised in the liver, exerts its effect on the mobilisation of Fe from liver stores by oxidation of the ferrous to ferric forms. Interrelationships between Cu and Mo on Fe metabolism are discussed by SEELIG [314]. The role of Cu in phospholipid synthesis has been closely investigated by GALLAGHER and co-workers [106–108] who established the necessity for Cu in the binding of adenine nucleotides to the mitochondrial membrane.

Cu is also associated with the functions of other enzymes, notably uricase, polyphenol oxidase and mono- and di-amine oxidases in various tissues.

Fluorine and *Iodine* when in soluble form are rapidly absorbed and, in common with chloride, are swiftly distributed throughout the tissues. In the case of F, that which is not incorporated into bones and teeth is eliminated mainly in the urine. Iodine excretion also follows the urinary path.

Mn absorption mechanisms are little known. Only a small amount of dietary Mn appears to be absorbed, and it is quickly excreted via the bile [116]. Mn disappearance from the bloodstream has been investigated by BORG and COTZIAS [28, 54] who showed that this occurred in three phases, one associated with normal small ion movement and the second and third associated with mitochondrial and nuclear turnover. An exponential analysis of the pattern of release of ^{54}Mn in the rhesus monkey [64] showed the presence of fast and slow components for the whole body and viscera. Mn-retention in the central nervous system appeared high.

Effects of differing levels of dietary Ca and phosphate on Mn absorption and retention have been reported in rats and chicks [173, 290].

The role of Mn in the body is associated with the activation of a wide variety of enzymes, often as an alternative to Mg. Specific roles for Mn were revealed by LEACH *et al.* [174] in chondroitin sulphate synthesis. The only known Mn metalloenzyme is pyruvate carboxylase [212, 313] but the essentiality of Mn for functioning of this enzyme, however, is doubtful in view of the findings of SCRUTTON *et al.* [313] that in Mn-depleted chicks the Mg-metalloenzyme form is found which retains catalytic activity.

Inorganic sulphate has marked effects upon *Mo* absorption and retention. Absorbed Mo is largely excreted in the urine, but the pattern of absorption and release from the body is affected by sulphate levels, whether of inorganic sulphate in the diet or that of endogenous origin derived from dietary protein. DICK [72] has suggested inorganic sulphate interferes with or prevents the transport of Mo through membranes. A rise in sulphate concentration would hinder reabsorption of Mo through the kidney tubule. In the chick, however, no lowering of tissue Mo concentration is effected by dietary sulphate [67, 68] even though the effects of Mo toxicity may be alleviated. Xanthine oxidase [32] and sulphite oxidase [51, 162] are Mo-metalloenzymes.

The elements Ni, Si, Sn and V are all poorly absorbed from the intestinal tract and the major excretory route is via the faeces.

Injection of ^{63}Ni into rats showed that the isotope was cleared almost completely from the body after 72 h [323]; only the kidney then contained significant amounts. This is in accord with other studies showing that Ni is not readily retained.

Poor retention of ingested *Si* in sheep is indicated by the work of JONES and HANDRECK [159]. Although urinary Si excretion rose with silica level in the diet, blood levels did not change, suggesting that the rate of absorption from the diet was not in excess of the ability of the kidneys to remove silica. Total faecal and urinary excretion amounted to 99% of the amounts ingested. Absorption of Si may be as monosilicic acid or finely divided silica.

Information on the metabolism of *Sn* is scanty. Excretion studies in human adults suggest that urinary output is less than 24 μg daily [253]. This compares with an estimated intake of from 1 to 38 mg daily according to the proportion of the diet derived from canned foods [295].

Although *V* is poorly absorbed, of that which is, a considerable proportion may be eliminated via the urine [74]; other studies [127, 328] show that absorbed V is retained for the most part in bones and teeth.

The essential biological roles of Si, V and Sn are, at present, unknown.

SCHWARZ [304], however, has described the occurrence of a bound form of silicon in the polysaccharide matrix of connective tissue.

Se absorption is rapid, although retention and distribution varies with the chemical form, dietary level and species of animal. Excretion patterns are also affected by dietary intake [139] and, in ruminants, a considerable proportion of ingested Se is eliminated in the faeces as an insoluble form, conversion presumably having occurred in the rumen [40, 55, 372]. Having entered the plasma, Se is carried to all tissues [38, 229], from where it may be lost in faeces, urine and in expired air [109].

The function of Se in the animal body in relation to vitamin E was initially believed to be that of a non-specific antioxidant which protected against peroxidation in tissues, and consequent disruption of membranes. DIPLOCK *et al.* [77], however, have shown that lipid peroxidation was not significant in vitamin E-deficient rats [76]. Further studies showed that Se existed in the tissues in three oxidation states and that the function of tocopherol may be to preserve the active selenide form of Se [46, 77], and that this form exists in a non-haem Fe-containing protein. The demonstration that Se appeared to be involved in the utilization of reduced glutathione in maintaining cell integrity, was followed by the finding that Se was an active component of glutathione peroxidase [278, 279]. The situation appears to be, therefore, that the enzymic function of Se is to promote peroxide destruction and that vitamin E by its solubility in the lipids of the membrane, prevents autoxidation of membrane lipids in a situation of marginal Se supply.

Zn absorption is influenced by other dietary factors, notably the presence of phytate and calcium [351]. Absorption in the rat occurs mostly in the duodenum by a carrier-mediated process [66] and recent studies by DAVIES [personal comm.] has shown a marked effect, in the rat, of the stage of pregnancy and lactation in increasing the binding to the mucosa and transport of Zn in the duodenum.

Absorbed Zn is incorporated into every tissue at differing rates [90, 128]. In the bones and the central nervous system, Zn exchange is slow. Urinary excretion of Zn accounts for only a small part of that excreted in the normal animal, and faecal Zn represents that lost from pancreatic secretions and shedding of intestinal mucosa in addition to any dietary Zn not absorbed.

In human subjects, a marked loss of urinary Zn occurs, notably in cases of nephrosis, alcoholic cirrhosis and hepatic porphyria [254, 339]. Total starvation in man also results in greatly increased urinary Zn output [330] as does tissue damage and surgical trauma [65, 94].

Zinc is a component of many enzymes, notably carbonic anhydrase, various dehydrogenases, pancreatic carboxypeptidase, DNA polymerase and alkaline phosphatase [247, 321]. There is no clear evidence that changes in the activity of these enzymes are related to the development of clinical lesions of Zn deficiency. The normal healthy animal is capable of holding tenaciously to its body Zn during periods of depletion and tissue Zn concentrations generally remain high if not, in many cases normal. Aberrations in Zn metabolism, however, have been noted in clinical conditions other than when gross tissue damage is apparent. An increased excretion of Zn, and less of Cu, was observed by PFEIFFER [255] in schizophrenic patients; the sedative effects of oral Zn suggested that these poeple had low brain Zn levels. A suggestion that Zn may be involved in the maturation and function of the hippocampal region of the brain derives from the observations of CRAWFORD and CONNOR [57].

Relations between Zn and stress factors are suggested by the observations of FLYNN *et al.* [96] that a strong positive correlation existed between serum Zn level and the rise in adrenocorticotrophic hormone during stress.

Taste acuity in patients in which body Zn is decreased (malabsorptive defects, Cushing's syndrome and multiple myeloma) was increased by treatment with Zn [131]. In this connexion it has been shown that Zn-deficient rats will preferentially select a Zn-containing diet, the response being noted within 4 h following the offering of a dietary choice [140].

The diversity of physiological or therapeutic effects attributable to Zn is exemplified further by the observations of DEMERTZIS and MILLS [70] who showed a stimulatory effect of oral Zn supplementation on the healing of pododermatitis in cattle. Relations between Zn and pathogenicity are supported by the work of SMITH *et al.* [322] who showed that germ-free rats had a lower Zn requirement than conventional animals; *Streptococcus* spp. and *Staphylococcus epidermidis* were identified as affecting Zn metabolism.

A role for Zn in the utilization and storage of vitamin A is suggested by a number of studies [193, 288, 324]. Liver folic acid stores appeared to be dependent on adequate Zn nutrition [370].

Effects of Zn on carbohydrate metabolism were shown by QUARTERMAN and co-workers [267–269].

The wide spectrum of defects in the living animal which arise as a consequence of the Zn deficiency state make it unlikely that they can be attributed to any fundamental metabolic lesion so far reported. The initial observation by LIEBERMAN and co-workers [184, 185] that the synthesis of

DNA is dependent on the presence of Zn has stimulated a considerable amount of research on the relation between Zn status and polynucleotide metabolism [48, 286, 338, 346, 361, 366, 369] with amply confirmed results showing diminutions in the rate of entry of labelled precursors into DNA of various tissues. The swift response of this process, in the whole animal, to depletion of Zn is paralleled by reductions in the activity of certain enzymes, notably carboxypeptidase and alkaline phosphatase of the gut [226, 365] which were shown to decrease in activity before growth and food intake were affected. It is clear, therefore, that localised Zn depletion can occur in cells even before the inhibition of DNA synthesis is sufficient to adversely affect growth. During the early stages of Zn depletion no adverse effects on testicular DNA synthesis were observed [366], thus illustrating a tissue-specific response. Such a response was also observed by FELL *et al.* [91] in a more advanced Zn depletion state in the rat. The apparently paradoxical observation that mitotic indices may increase in some tissues and decrease or be unaffected in others by the Zn deficiency state implies that a generalised decrease in the rate of DNA synthesis cannot be held responsible for the decline in growth of all tissues. HURLEY and co-workers [346] noted, however, that increased numbers of mitotic figures in the neuroepithelium of Zn-deficient rat embryos as compared with controls was not associated with ^{3}H-thymidine incorporation into these cells. These findings suggest that Zn may be associated with final stages in the maturation of cells. If this were so, it appears likely that reductions in the rate of incorporation of other labelled precursors into protein [150–152; 201] may be a reflection of the failure of cells to mature rather than a cause of that effect. This situation remains unresolved. Zinc metabolism and function still present many formidable problems to be solved; the role of Zn as a stabiliser of macromolecules is reviewed by CHVAPIL [50].

It would be inappropriate to conclude this review without some mention of the vital necessity, in animal experimentation, of paying particular attention to dietary trace nutrient requirements. It is, unfortunately, a common experience among workers in the field of trace elements to read papers derived from other forms of biological investigation in which it is apparent that only the scantiest attention, if any, has been paid to the adequacy or otherwise of mineral supplementation, even when it is obvious that deficiencies or excesses of one element or other are known to affect the processes being investigated. In this regard, the reader's attention is drawn to the excellent review by GREENFIELD and BRIGGS [117] in which nutritional methodology in metabolic research is evaluated.

Summary

Trace elements are considered as existing as inorganic ion-organic ligand complexes, competition between them being responsible for manifestations of depletion or toxicity states, in addition to their normal role in biological processes. Within the limitations of space imposed, it has not been possible to examine in detail many of the interactions known to occur. Particular attention has been paid to the manifestation of disease states in animal and human conditions, especially in regard to the elements Cu, Se and Zn. That many gaps exist in knowledge of basic requirements for optimum physiological performance has been pointed out, as well as the fact that for many species (both vertebrate and invertebrate) insufficient knowledge of trace element requirement or metabolism has yet accrued to permit reliable prediction of the development of deficiency or toxicity states within a given set of environmental circumstances. Though present in only a small quantity in tissues, trace elements have unique and vital roles in the control of metabolic processes. Regrettably, however, methods available for the identification of deficiency or toxicity states in animals leave much to be desired, with the consequence that their impact is frequently overlooked. This article, it is hoped, may serve to emphasise the need for wider consideration of the roles and importance of trace elements in animal populations.

References

1 Abdel-Salam, E.; Allah, A. K., and Osman, N. G.: Anaemia of protein malnutrition. Acta biol. med. germ. *27:* 279–288 (1971).

2 Akey, D. H. and Beck, S. D.: Nutrition of the pea aphid *Acyrthosiphon pisum.* Requirements for trace metals, sulphur and cholesterol. J. insect Physiol. *18:* 1901–1914 (1972).

3 Allaway, J. H. and Hodgson, J. F.: Symposium on nutrition, forage and pastures. Selenium in forages as related to the geographic distribution of muscular dystrophy in livestock. J. anim. Sci. *23:* 271–277 (1964).

4 Ammerman, C. B.; Arrington, L. R.; Warnick, A. C.; Edwards, J. L.; Shirley, R. L., and Davis, G. K.: Reproduction and lactation in rats fed excessive iodine. J. Nutr. *84:* 107–112 (1964).

5 Anke, M. and Groppel, B.: Manganese deficiency and radioisotope studies on manganese metabolism; in Mills Proc. 1st Int. Symp. Trace Element Metab. Anim., Aberdeen 1969, pp. 133–136 (Livingstone, Edinburgh 1970).

6 Anke, M.; Groppel, B.; Reissig, W.; Luedke, H.; Gruen, M. und Dittrich, G.: Manganmangel beim Wiederkäuer. 3. Manganmangelbedingte Fortpflanzungs-Skelett- und Nervenstörungen bei weiblichen Wiederkäuern und ihren Nachkommen. Arch. Tierernähg. *23:* 197–211 (1973).

7 Apgar, J.: Effect of zinc deficiency on parturition in the rat. Am. J. Physiol. *215:* 160–163 (1968).

8 Apgar, J.: Effect of a low-zinc diet during gestation on reproduction in the rabbit. J. Anim. Sci. *33:* 1255–1258 (1971).

9 APGAR, J.: Effect of zinc deprivation from day 12, 15 or 18 of gestation on parturition in the rat. J. Nutr. *102:* 343–348 (1972).

10 ARRINGTON, L. R.; SANTA CRUZ, R. A.; HARMS, R. H., and WILSON, H. R.: Effects of excess dietary iodine upon pullets and laying hens. J. Nutr. *92:* 325–333 (1967).

11 ARRINGTON, L. R.; TAYLOR, R. N.; AMMERMAN, C. B., and SHIRLEY, R. L.: Effects of excess iodine upon rabbits, hamsters, rats and swine. J. Nutr. *87:* 394–398 (1965).

12 ASHIDA, M.: Purification and characteristics of pre-phenoloxidase from haemolymph of the silkworm, *Bombyx mori.* Archs Biochem. Biophys. *144:* 749–762 (1971).

13 ATKINSON, R. L.; BRADLEY, J. W.; COUCH, J. R., and QUISENBERRY, J. H.: Effects of various levels of manganese on the reproductive performance of turkeys. Poultry Sci. *46:* 472–475 (1967).

14 AUCLAIR, J. L. and SRIVASTAVA, P. N.: Some mineral requirements of the pea aphid *(Acyrthosiphon pisum).* Can. Entomol. *104:* 927–936 (1972).

15 BALINSKY, J. B.: Nitrogen metabolism in amphibians; in CAMPBELL Comparative biochemistry of nitrogen metabolism, vol. 2, pp. 519–637 (Academic Press London 1970).

16 BARLOW, R. M.: Further observations on swayback. I. Transitional pathology. J. comp. Path. *73:* 51–60 (1963).

17 BARLOW, R. M.: Further observations on swayback. II. Histochemical localisation of cytochrome oxidase activity in the central nervous system. J. comp. Path. *73:* 61–67 (1963).

18 BARNEY, G. H.; MACAPINLAC, M. P.; PEARSON, W. N., and DARBY, W. J.: Parakeratosis of the tongue. A unique histopathologic lesion in the zinc-deficient squirrel monkey. J. Nutr. *93:* 511–522 (1967).

19 BARNEY, G. H.; ORGEBIN-CRIST, M.-C., and MACAPINLAC, M. P.: Oesophageal parakeratosis and histologic changes in the testes of the zinc-deficient rat and their reversal by zinc repletion. J. Nutr. *95:* 526–534 (1968).

20 BARR, D. H. and HARRIS, J. W.: Growth of the P388 leukemia as an ascites tumour in zinc-deficient mice. Proc. Soc. exp. Biol. Med. *144:* 284–287 (1973).

21 BELLISARIO, R. and CORMIER, M. J.: Peroxide-linked bioluminescence catalysed by a copper-containing non-haeme luciferase isolated from a bioluminescent earthworm. Biochem. biophys. Res. Commun. *43:* 800–805 (1971).

22 BECKER, D. E. and SMITH, S. E.: The level of cobalt tolerance in yearling sheep. J. Anim. Sci. *10:* 266–271 (1951).

23 BENNETS, H. W. and BECK, A. B.: in UNDERWOOD Trace elements in human and animal nutrition; 3rd ed. (Academic Press, London 1971).

24 BENTLEY, O. G. and PHILLIPS, P. H.: Effect of low manganese rations upon dairy cattle. J. Dairy Sci. *34:* 396–403 (1951).

25 BIRCKNER, V.: The zinc content of some foodstuffs. J. biol. Chem. *38:* 191–203 (1919).

26 BLAXTER, K. L.: The effect of selenium administration on the growth and health of sheep on Scottish farms. Br. J. Nutr. *17:* 105–115 (1963).

27 BLOKHINA, R. I.: Geochemical ecology of endemic goitre; in MILLS Proc. 1st Int. Symp. Trace Element Metab. Anim., Aberdeen 1969, pp. 426–432 (Livingstone, Edinburgh 1970).

28 BORG, D. C. and COTZIAS, G. G.: Manganese metabolism in man. Rapid exchange

of Mn^{56} with tissue as demonstrated by blood clearance and liver uptake. J. clin. Invest. *37:* 1269–1278 (1958).

29 Boyden, R.; Potter, V. R., and Elvehjem, C. A.: Effect of feeding high levels of copper to albino rats. J. Nutr. *15:* 397–402 (1938).

30 Boyer, P. D.; Shaw, J. H., and Phillips, P. H.: Studies on manganese deficiency in the rat. J. biol. Chem. *143:* 417–425 (1942).

31 Braude, R.; Free, A. A.; Page, J. E., and Smith, E. L.: The distribution of radioactive cobalt in pigs. Br. J. Nutr. *3:* 289–292 (1949).

32 Bray, R. C.; Petersson, R., and Ehrenberg, A.: The chemistry of xanthine oxidase. 7. The anaerobic reduction of xanthine oxidase studied by electron-spin resonance and magnetic susceptibility. Biochem. J. *81:* 178–189 (1956).

33 Bremner, I.: Zinc, copper and manganese in the alimentary tract of sheep. Br. J. Nutr. *24:* 769–783 (1970).

34 Bremner, I. and Knight, A. H.: The complexes of zinc, copper and manganese present in rye grass. Br. J. Nutr. *24:* 279–289 (1970).

35 Brooks, M. A.: Some dietary factors that affect ovarial transmission of symbiotes. Proc. helminth. Soc. Wash. *27:* 212–220 (1960).

36 Bryan, G. W.: Zinc concentration of fast and slow contracting muscles in the lobster. Nature, Lond. *213:* 1043–1044 (1967).

37 Buchanan-Smith, J. G.; Nelson, E. C.; Osburn, B. I.; Wells, M. E., and Tillman, A. D.: Effects of vitamin E and selenium deficiencies in sheep fed a purified diet during growth and reproduction. J. Anim. Sci. *29:* 808–815 (1969).

38 Buescher, R. G.; Bell, M. C., and Berry, R. K.: Effect of excessive calcium on selenium-75 in swine (abstr.) J. Anim. Sci. *19:* 1251–1252 (1960).

39 Burk, R. F.; Whitney, R.; Frank, H., and Pearson, W. W.: Tissue selenium levels during the development of dietary liver necrosis in rats fed *Torula* yeast diets. J. Nutr. *95:* 420–428 (1968).

40 Butler, G. W. and Peterson, P. J.: Aspects of the faecal excretion of selenium by sheep. N. Z. J. agric. Res. *4:* 484–491 (1961).

41 Caldwell, D. F.; Oberleas, D., and Prasad, A. S.: Reproductive performance of chronic mildly zinc-deficient rats and the effects on behaviour of their offspring. Nutr. Rep. Int. *7:* 309–319 (1973).

42 Carlisle, E. M.: Silicon. Essential element for the chick. Science *178:* 619–621 (1972).

43 Carnes, W. H.; Shields, G. S.; Cartwright, G. E., and Wintrobe, M. M.: Vascular lesions in copper deficient swine (abstr.). Fed. Proc. Fed. Am. Socs exp. Biol. *20:* 118 (1961).

44 Cartwright, G. F. and Wintrobe, M. M.: Question of copper deficiency in man. Am. J. clin. Nutr. *15:* 94–110 (1964).

45 Cassens, R. G.; Briskey, E. J., and Hoekstra, W. G.: Variation in zinc content and other properties of various porcine muscles. J. Sci. Food Agric. *14:* 427–432 (1963).

46 Caygill, C. P. J.; Lucy, J. A., and Diplock, A. T.: Effect of vitamin E on the intracellular distribution of the different oxidation states of selenium in rat liver. Biochem. J. *125:* 407–416 (1971).

47 Chatterjee, G. C.; Roy, R. K.; Sasmal, N.; Banerjee, S. K., and Majumder, P. K.:

Effect of chromium and tungsten on L-ascorbic acid metabolism in rats. J. Nutr. *103:* 509–514 (1973).
48 CHESTERS, J. K.: The role of zinc ions in the transformation of lymphocytes by phytohaemagglutinin. Biochem. J. *130:* 133–139 (1972).
49 CHESTERS, J. K. and QUARTERMAN, J.: Effects of zinc deficiency on food intake and feeding patterns of rats. Br. J. Nutr. *24:* 1061–1069 (1970).
50 CHVAPIL, M.: New aspects in the biological role of zinc. A stabiliser of macromolecules and biological membranes. Life Sci. *13:* 1041–1049 (1973).
51 COHEN, H. J.; FRIDOVICH, I., and RAJAGOPALAN, K. V.: Hepatic sulphite oxidase. Functional role for molybdenum. J. biol. Chem. *246:* 374–382 (1971).
52 CORDANO, A.; BAERTL, J. M., and GRAHAM, G. G.: Copper deficiency in infancy. Pediatrics *34:* 325–336 (1964).
53 CORDANO, A.; PLACKO, R. P., and GRAHAM, G. G.: Hypocupremia and neutropenia in copper deficiency. Blood *28:* 280–283 (1966).
54 COTZIAS, G. G.: Manganese in health and disease. Physiol. Rev. *38:* 503–532 (1958).
55 COUSINS, F. B. and CAIRNEY, I. M.: Some aspects of selenium metabolism in sheep. Austr. J. agric. Res. *12:* 927–943 (1961).
56 COX. D. H. and HARRIS, D. L.: Effect of excess dietary zinc on iron and copper in the rat. J. Nutr. *70:* 514–520 (1960).
57 CRAWFORD, I. L. and CONNOR, J. D.: Zinc in maturing rat brain: Hippocampal concentration and localisation. J. Neurochem. *19:* 1451–1458 (1972).
58 CRESS, D. C. and CHADA, H. L.: Development of a synthetic diet for the greenbug, *Schizaphis graminum.* 2. Greenbug development as affected by zinc, iron, manganese and copper. Ann. Entomol. Soc. Am. *64:* 1240–1244 (1971).
59 CUNNINGHAM, I. J.: Some biochemical and physiological aspects of copper in animal nutrition. Biochem. J. *25:* 1267–1294 (1931).
60 CUNNINGHAM I. J.: in UNDERWOOD Trace elements in human and animal nutrition; 3rd ed. (Academic Press, London 1971).
61 CVANCARA, V.A.: Liver arginase activity in the sockeye salmon, *Oncorrhyncus nerka.* Comp. Biochem. Physiol. B *40:* 819–822 (1971).
62 DADD, R. H.: Insect nutrition. Current developments and metabolic implications; in Annu. Rev. Entomol., pp. 381–420 (Annual Reviews, Palo Alto, 1973).
63 DANKS, D. M.; CARTWRIGHT, E.; STEVENS, B. J., and TOWNLEY, R. R. W.: Menkes kinky hair disease. Further definition of the defect in copper transport. Science *179:* 1140–1142 (1973).
64 DASTUR, D. K.; MANGHANI, D. K., and RAGHAVENDRA, K. V.: Distribution and fate of manganese-54 in the monkey. Studies of different parts of the central nervous system and other organs. J. clin. Invest. *50:* 9–20 (1971).
65 DAVIES, J.W. L. and FELL, G. S.: Tissue catabolism in patients with burns. Clin. chim. Acta *51:* 83–92 (1974).
66 DAVIES, N. T.: Intestinal transport of zinc by rat duodenum. J. Physiol. *229:* 46–47P (1972).
67 DAVIES, R. E.; REID, B. L., and COUCH, J. R.: Storage of molybdenum and copper in livers of chicks (abstr.) Fed. Proc. Fed. Am. Socs exp. Biol. *18:* 522 (1959).
68 DAVIES, R. E.; REID, B. L.; KURNICK, A.A., and COUCH, J. R.: The effect of sulphate on molybdenum toxicity in the chick. J. Nutr. *70:* 193–198 (1960).

69 JORGE, F. B., DE; HAESER, P. E.; DITADI, A. S. F.; PETERSEN, J. A.; ULHÔA CINTRA, A. B., and SAWAYA, P.: Biochemical studies on the giant earthworm *Glossoscolex giganteus* (Leuckart). Comp. Biochem. Physiol. *16:* 491–496 (1965).

70 DEMERTZIS, P. and MILLS, C. F.: Oral zinc therapy in the control of infectious pododermatitis in young bulls. Vet. Rec. *93:* 219–222 (1973).

71 DEWYS, W. and PORIES, W. J.: Inhibition of a spectrum of animal tumours by dietary zinc deficiency. J. natn. Cancer Inst. *48:* 375–381 (1972).

72 DICK, A. T.: in UNDERWOOD Trace elements in human and animal nutrition; 3rd ed. (Academic Press, London 1971).

73 DICKSON, R. C. and TOMLINSON, R. H.: Selenium in blood and human tissues. Clin. chim. Acta *16:* 311–321 (1967).

74 DIMOND, E. G.; CARAVACA, J., and BENCHIMAL, A.: Vanadium. Excretion, toxicity, lipid effect in man. Am. J. clin. Nutr. *12:* 49–53 (1963).

75 DINKEL, C. A.; MINYARD, J. A., and RAY, D. E.: Effect of season of breeding on reproduction and weaning performance of beef cattle grazing seleniferous range. J. Anim. Sci. *22:* 1043–1045 (1963).

76 DIPLOCK, A. T.: Recent studies on the interactions between vitamin E and selenium; in MILLS Proc. 1st Int. Symp. Trace Element Metab. Anim., Aberdeen 1969, pp. 190–204 (Livingstone, Edinburgh 1970).

77 DIPLOCK, A. T.; BAUM, H., and LUCY, J. A.: Effect of vitamin E on the oxidation state of selenium in rat liver. Biochem. J. *123:* 721–729 (1971).

78 DOISY, R. J.; STREETEN, D. H. P.; SOUMA, M. L.; KALAFER, M. E.; REKANT, S. I., and DALAKOS, T. G.: Metabolism of ^{51}chromium in human subjects. I. Normal, elderly and diabetic subjects; in MERTZ and CORNATZER Newer trace metals in nutrition, pp. 155–168 (Marcel Dekker, New York 1971).

79 DREOSTI, I. E.; GREY, P. C., and WILKINS, P. J.: Deoxyribonucleic acid synthesis, protein synthesis and teratogenesis in zinc-deficient rats. S. Afr. med. J. *46:* 1585–1588 (1972).

80 DREOSTI, I. E.; TAO, S., and HURLEY, L. S.: Plasma zinc and leucocyte changes in weanling and pregnant rats during zinc deficiency. Proc. Soc. exp. Biol. Med. *128:* 169–174 (1968).

81 DUTT, B. and MILLS, C. F.: Reproductive failure in rats due to copper deficiency. J. comp. Path. *70:* 120–125 (1960).

82 EGGERT. R. G.; PATTERSON, E. L.; AKERS, W. T., and STOKSTAD, E. L. R.: The role of vitamin E and selenium in the nutrition of the pig (abstr.) J. Anim. Sci. *16:* 1037 (1957).

83 ELLIS, W. C.; PFANDER, W. H.; MUHRER, M. E., and PICKETT, E. E.: Molybdenum as a dietary essential for lambs. J. Anim. Sci. *17:* 180–188 (1958).

84 ERWAY, L.; FRASER, A. S., and HURLEY, L. S.: Prevention of congenital otolith defect in *pallid* mutant mice by manganese supplementation. Genetics *67:* 97–108 (1971).

85 ERWAY, L.; HURLEY, L. S., and FRASER, A. S.: Neurological defect. Manganese in phenocopy and prevention of a genetic abnormality of inner ear. Science *152:* 1766–1768 (1966).

86 ERWAY, L.; HURLEY, L. S., and FRASER, A. S.: Congenital ataxia and otolith defects due to manganese deficiency in mice. J. Nutr. *100:* 643–654 (1970).

87 ESTABLIER, R.: Copper, zinc, manganese and iron content of ovaries of tuna, *Thunnus thynnus;* bonito, *Sarda sarda;* bacoreta, *Euthynnus alletteratus,* and melva, *Auxis thazard.* Invest. Pesq. *34:* 171–175 (1970); cited in Chem. Abstr. *75:* 1852 (1971).

88 EVERSON, G. J.; HURLEY, L. S., and GEIGER, J. F.: Manganese deficiency in the guinea-pig. J. Nutr. *68:* 49–56 (1959).

89 EVERSON, G. J.; TSAI, M.-C., and WANG, T.: Copper deficiency in the guinea-pig. J. Nutr. *93:* 533–540 (1967).

90 FEASTER, J. P.; HANSARD, S. L.; MCCALL, J. T.; SKIPPER, F. H., and DAVIS, G. K.: Absorption and tissue distribution of radiozinc in steers fed high-zinc rations. J. Anim. Sci. *13:* 781–788 (1954).

91 FELL, B. F.; LEIGH, L. C., and WILLIAMS, R. B.: The cytology of various organs in zinc-deficient rats with particular reference to the frequency of cell division. Res. vet. Sci. *14:* 317–325 (1973).

92 FELL, B. F.; MILLS, C. F., and BOYNE, R.: Cytochrome oxidase deficiency in the motor neurones of copper-deficient lambs. A histochemical study. Res. vet. Sci. *6:* 170–177 (1965).

93 FELL, B. F.; WILLIAMS, R. B., and MILLS, C. F.: Further studies of nervous tissue degeneration resulting from conditioned copper deficiency in lambs. Proc. Nutr. Soc. *20:* xxvii (1961).

94 FELL, G. S.; CUTHBERTSON, D. P.; FLECK, A.; QUEEN, K.; MORRISON, C.; BESSENT, R. G., and HUSAIN, S. L.: Urinary zinc levels as an indication of muscle catabolism. Lancet *i:* 280–282 (1973).

95 FERGUSON, W. S.: The teart pastures of Somerset. IV. The effect of continuous administration of copper sulphate to dairy cows. J. agric. Sci. *33:* 116–118 (1943).

96 FLYNN, A.; PORIES, W. J.; STRAIN, W. H., and HILL, O. A.: Mineral element correlations with adeno-hypophyseal-adrenal cortex function and stress. Science *173:* 1035–1036 (1971).

97 FOLLIS, R. H.: Deficiency disease, pp. 71–72 (Ch. C. Thomas, Springfield 1958).

98 FORE, H. and MORTON, R. A.: The manganese in bone. Biochem. J. *51:* 598–600 (1952).

99 FORE, H. and MORTON, R. A.: Manganese in rabbit tissues. Biochem. J. *51:* 600–603 (1952).

100 FORE, H. and MORTON, R. A.: Manganese in eye tissues. Biochem. J. *51:* 603–606 (1952).

101 FRANKEL, G.: The effect of zinc and potassium in the nutrition of *Tenebrio molitor* with observations on the expression of a carnitine deficiency. J. Nutr. *65:* 361–395 (1958).

102 FRANKE, K. W. and MOXON, A. L.: The toxicity of orally ingested arsenic, selenium, tellurium, vanadium and molybdenum. J. Pharmacol. exp. Ther. *61:* 89–101 (1937).

103 FRANKE, K. W. and POTTER, V. R.: A new toxicant occurring naturally in certain samples of plant foodstuffs. J. Nutr. *10:* 213–221 (1935).

104 FRANKE, K. W. and TULLY, W. C.: A new toxicant occurring naturally in certain samples of plant foodstuffs. Poultry Sci. *14:* 273–279 (1935).

105 FRANKE, K. W. and TULLY, W. C.: A new toxicant occurring naturally in certain samples of plant foodstuffs. VII. Low hatchability due to deformities in chicks

produced from eggs obtained from chickens of known history. Poultry Sci. *15:* 316–318 (1936).
106 GALLAGHER, C. H. and REEVE, V. E.: Copper deficiency in the rat. Effect on synthesis of phospholipids. Aust. J. exp. Biol. med. Sci. *49:* 21–31 (1971).
107 GALLAGHER, C. H. and REEVE, V. E.: Copper deficiency in the rat. Effect on adenine nucleotide metabolism. Aust. J. exp. Biol. med. Sci. *49:* 445–451 (1971).
108 GALLAGHER, C. H.; REEVE, V. E., and WRIGHT, R.: Copper deficiency in the rat. Effect on the ultrastructure of hepatocytes. Aust. J. exp. Biol. med. Sci. *51:* 189–197 (1973).
109 GANTHER, H. E.; LEVANDER, O. A., and BAUMANN, C. A.: Dietary control of selenium volatilisation in the rat. J. Nutr. *88:* 55–60 (1966).
110 GODFREY, P. R.; CARRICK, C. W., and QUACKENBUSH, F. W.: Iodine nutrition of chicks. Poultry Sci. *32:* 394–396 (1953).
111 GOLLAN, J. L.; DAVIS, P. S., and DELLER, D. J.: Binding of copper by human alimentary secretions. Am. J. clin. Nutr. *24:* 1025–1027 (1971).
112 GOLLAN, J. L. and DELLER, D. J.: Nature and excrection of biliary copper in man. Clin. Sci. *44:* 9–15 (1973).
113 GOPINATH, C.; HALL, G. A., and HOWELL, J. MCC.: The Effects of chronic copper poisoning on the kidneys of sheep. Res. vet. Sci. *16:* 57–69 (1974).
114 GORDON, H. T.: Minimal nutritional requirements of the German roach, *Blatella germanica L.* Ann. N.Y. Acad. Sci. *77:* 290–351 (1959).
115 GRAHAM, G. G. and CORDANO, A.: Copper depletion and deficiency in the malnourished infant. Johns Hopkins med. J. *124:* 139–150 (1969).
116 GREENBERG, D. M.; COPP, H. D., and CUTHBERTSON, E. M.: Studies in mineral metabolism with the aid of artificial radioactive isotopes. VII. The distribution and excretion, particularly by way of the bile of iron, cobalt and manganese. J. biol. Chem. *147:* 749–756 (1943).
117 GREENFIELD, H. and BRIGGS, G. M.: Nutritional methodology in metabolic research with rats. Annu. Rev. Biochem. *40:* 549–572 (1971).
118 GRIES, C. L. and SCOTT, M. L.: Pathology of selenium deficiency in the chick. J. Nutr. *102:* 1287–1296 (1972).
119 GROPPEL, B.; ANKE, M.; HAHN, G. und BENSER, A.: Manganmangel beim Wiederkäuer. 2. Mitteilung: Der Einfluß der Manganversorgung auf die Fortpflanzungsleistung und Ejakulatzusammensetzung. Arch. exp. Vet. Med. *27:* 383–386 (1973).
120 GRUMMER, R. H.; BENTLEY, O. G.; PHILLIPS, P, H., and BOHSTEDT, G.: The role of manganese in growth, reproduction and lactation of swine. J. Anim. Sci. *9:* 170–175 (1950).
121 HALVERSON, A.W. and MONTY, K. J.: An effect of dietary sulphate on selenium poisoning in the rat. J. Nutr. *70:* 100–102 (1960).
122 HAMBIDGE. K. M.: Chromium nutrition in the mother and growing child; in MERTZ and CORNATZER Newer trace metals in nutrition, pp. 169–194 (Marcel Dekker, New York 1971).
123 HARPER, R. A. and ARMSTRONG, F. B.: Alkaline phosphatase of *Drosophila melanogaster*. I. Partial purification and characterisation. Biochem. Genet. *6:* 75–82 (1972).
124 HARTLEY, W. J.; DRAKE, C., and GRANT, A. B.: Selenium and animal health. N.Z. J. Agric. *99:* 259–264 (1959).

125 HARTLEY, W. J. and GRANT, A. B.: A review of selenium-responsive diseases in New Zealand livestock. Fed. Proc. Fed. Am. Socs exp. Biol. *20:* 679–688 (1961).

126 HARTMANS, J.: in UNDERWOOD Trace elements in human and animal nutrition; 3rd ed. (Academic Press, London 1971).

127 HATHCOCK, J. N.; HILL, C. H., and MATRONE, G.: Vanadium toxicity and distribution in chicks and rats. J. Nutr. *82:* 106–110 (1964).

128 HEATH, J. C. and LIQUIER-MILWARD, J.: The distribution and function of zinc in normal and malignant tissues. I. Uptake and distribution of radioactive zinc, ^{65}Zn. Biochim. biophys. Acta *5:* 404–415 (1950).

129 HEDGES, R. S.; HOWARD, D. A., and BURDIN, M. L.: The occurrence in goats and sheep in Kenya of a disease closely similar to swayback. Vet. Rec. *76:* 493–497 (1964).

130 HELLER, V. G. and BURKE, A. D.: Toxicity of zinc. J. biol. Chem. *74:* 85–93 (1927).

131 HENKIN, R. I.: Trace metals and taste; in HEMPHILL Trace substances in environmental health. Proc. 3rd Univ. Mo. Ann. Conf. 1969, pp. 169–181 (Univ. of Missouri Press, Columbia, Mo. 1970).

132 HENNIG, A.; ANKE. M.; GROPPEL, B.; LUEDKE, H.; REISSIG, W.; DITTRICH, G. und GRUEN, M.: Manganmangel beim Wiederkäuer. 1. Der Einfluß des Manganmangels auf die Lebendmaßentwicklung. Arch. Tierernähg. *22:* 601–614 (1972).

133 HENRIKSON, P. A.; LUTWAK, L.; CROOK, L.; KALLFELZ, F.; SHEFFY, B. E.; SKOGERBOE, R.; BELANGER, L. F.; MARIER, J. R.; ROMANUS, B., and HIRSCH, C.: Fluoride and nutritional osteoporosis. Fluoride *3:* 204–207 (1970).

134 HIGGINS, E. S.; RICHERT, D. A., and WESTERFELD, W. W.: Molybdenum deficiency and tungstate inhibition studies. J. Nutr. *59:* 539–559 (1956).

135 HILL, R. and MATHER, J. W.: Manganese in the nutrition and metabolism of the pullet. I. Shell thickness and manganese content of eggs from birds given a diet of low or high manganese content. Br. J. Nutr. *22:* 625–633 (1968).

136 HOEKSTRA, W. G.; FALTIN, E. C.; LIN, C. W.; ROBERTS, H. F., and GRUMMER, R. H.: Zinc deficiency in reproducing gilts fed a diet high in calcium and its effect on tissue zinc and blood serum alkaline phosphatase. J. Anim. Sci. *26:* 1348–1457 (1967).

137 HOPKINS, L. L.: Distribution in the rat of physiological amounts of injected Cr^{51} (III) with time. Am. J. Physiol. *209:* 731–735 (1965).

138 HOPKINS, L. L., jr. and MOHR, H. E.: The biological essentiality of vanadium; in MERTZ and CORNATZER Newer trace elements in nutrition, pp. 195–213 (Marcel Dekker, New York 1971).

139 HOPKINS, L. L., jr.; POPE, A. L., and BAUMANN, C. A.: Distribution of microgram quantities of selenium in the tissues of the rat, and effects of previous selenium intake. J. Nutr. *88:* 61–65 (1966).

140 HUMPHRIES, W. R. and QUARTERMAN, J.: The selection of zinc-containing and protein-free diets by zinc-deficient rats. Proc. Nutr. Soc. *27:* 54A (1968).

141 HURLEY, L. S. and BELL, L. T.: Genetic influence on response to dietary manganese deficiency in mice. J. Nutr. *104:* 133–137 (1974).

142 HURLEY, L. S.; EVERSON, G. J., and GEIGER, J. F.: Manganese deficiency in rats. Congenital nature of ataxia. J. Nutr. *66:* 309–319 (1958).

143 HURLEY, L. S.; GOWAN, J., and SWENERTON, H.: Teratogenic effects of short-term and transitory zinc deficiency in rats. Teratology *4:* 199–204 (1971).

144 HURLEY, L. S. and MUTCH, P. B.: Prenatal and post natal development after transitory gestational zinc deficiency in rats. J. Nutr. *103:* 649–656 (1973).
145 HURLEY, L. S. and SWENERTON, H.: Congenital malformations resulting from zinc deficiency in the rat. Proc. Soc. exp. Biol. Med. *123:* 692–696 (1966).
146 HURT, H. D.; CARY, E. E., and VISEK, W. J.: Growth, reproduction and tissue concentration of selenium in the selenium-depleted rat. J. Nutr. *101:* 761–766 (1971).
147 HUTNER, S. H.: Inorganic nutrition. Annu. Rev. Microbiol. 26: 313–346 (1972).
148 HUTNER, S. H.; AARONSON, S.; NATHAN, H. A.; BAKER, H.; SCHER, S., and CURY, A.: Trace elements in microorganisms. The temperature factor approach; in LAMB *et al.* Trace elements, pp. 47–65 (Academic Press, London 1958).
149 HUTNER, S. H.; BAKER, H.; FRANK, C., and COX, D.: Nutrition and metabolism in protozoa; in FIENNES Biology of nutrition, pp. 85–177 (Pergamon Press, Oxford 1972).
150 HSU, J. M. and ANTHONY, W. L.: Zinc deficiency and collagen synthesis in rat skin; in HEMPHILL Trace substances in environmental health, vol. 6, pp. 137–143 (Univ. of Missouri Press, Columbia 1973).
151 HSU, J. M.; ANTHONY, W. L., and BUCHANAN, P. J.: Zinc deficiency and incorporation of ^{14}C-labelled methionine into tissue proteins in rats. J. Nutr. *99:* 425–432 (1969).
152 HSU, J. M. and WOOSLEY, R. L.: Metabolism of L-methionine-^{35}S in zinc-deficient rats. J. Nutr. *102:* 1181–1186 (1972).
153 IMAI, Y.: Relation between fluoride concentration in drinking water and dental caries in Japan. Koku Eisei Gakkai Zasshu *22:* 144–196 (1972); cited in Chem. Abstr. *78:* 82538 (1973).
154 IMAI, Y. and KITAMURU, T.: Dental caries and fluoride content of well water in Krosecho, Japan. Koku Eisei Gakkai Zasshu *22:* 285–294 (1972); cited in Chem. Abstr. *78:* 82539 (1973).
155 IMAI, Y. and OKADA, S.: Relation between fluoride concentration in drinking water and dental caries prevalence in Shodo island of Japan. Koku Eisei Gakkai Zasshu *22:* 269–280 (1972); cited in Chem. Abstr. *78:* 82540 (1973).
156 INNES, J. R. M. and SHEARER, G. D.: 'Swayback', a demyelinating disease of lambs with affinities to Schilder's encephalitis in man. J. comp. Path. *53:* 1–41 (1940).
157 JENSEN, L. S.: Selenium deficiency and impaired reproduction in Japanese quail. Proc. Soc. exp. Biol. Med. *128:* 970–972 (1968).
158 JOHNSON, S. R.: Studies with swine on rations extremely low in manganese. J. Anim. Sci. *2:* 14–22 (1943).
159 JONES, L. H. P. and HANDRECK, K. A.: The relation between the silica content of the diet and the excretion of silica by the sheep. J. agric. Sci. *65:* 129–134 (1965).
160 KEHOE, R. A.; CHOLAK, J., and STOREY, R. V.: A spectrochemical study of the normal ranges of concentration of certain trace metals in biological materials. J. Nutr. *19:* 579–592 (1940).
161 KEMMERER, A. R.; ELVEHJEM, C. A., and HART, E. B.: Studies on the relation of manganese to the nutrition of the mouse. J. biol. Chem. *92:* 623–630 (1931).
162 KESSLER, D. L. and RAJAGOPALAN, K. V.: Purification and properties of sulfite oxidase from chicken liver. Presence of molybdenum in sulfite oxidase from diverse sources. J. biol. Chem. *129:* 231–239 (1972).

163 KHODAK, A. A.: Levels of iron, copper and cholinesterase activity of the blood in women in the process of labour in uterine inertia. Akush. Ginekol, Moscow *47:* 69–71 (1971); cited in Chem. Abstr. *75:* 96624 (1971).

164 KIDDER, G. W.; DEWEY, V. C., and PARKS, R. E.: Studies on the inorganic requirements of *Tetrahymena.* Physiol. Zool. *24:* 69–75 (1951).

165 KIDDER, R. W.: Symptoms of induced copper toxicity in a steer (abstr.) J. Anim. Sci. *8:* 623–624 (1949).

166 KIENHOLZ, E. W.; TURK, D. E.; SUNDE, M. L., and HOEKSTRA, W. G.: Effects of zinc deficiency in the diets of hens. J. Nutr. *75:* 211–221 (1961).

167 KING, E. J. and BELT, T. H.; The physiological and pathological effects of silica. Physiol. Rev. *18:* 329–365 (1938).

168 KRATZER, F. H.; VOHRA, P.; ALLRED, J. B., and DAVIS, P. N.: Effect of zinc upon growth and incidence of perosis in turkey poults. Proc. Soc. exp. Biol. Med. *98:* 205–207 (1958).

169 KRISHNAMACHARI, K. A. V. R. and RAO, K. S. J.: Ceruloplasmin activity in infants born to malnourished mothers. J. Obstet. Gynaec. Br. Commonw. *79:* 162–165 (1972).

170 LAHEY, M. E.; BEHAR, M.; VITERI, F., and SCRIMSHAW, N. S.: Values for copper, iron and iron-binding capacity in the serum in Kwashiokor. Pediatrics *23:* 71–79 (1958).

171 LAHEY, M. E.; GUBLER, C. J.; CHASE, M. S.; CARTWRIGHT, G. E., and WINTROBE, M. M.: Studies on copper metabolism. II. Hematologic manifestations of copper deficiency in swine. Blood *7:* 1053–1074 (1952).

172 LANG, C. A.: Occurrence of zinc in mosquito and other species. Fed. Proc. Fed. Am. Socs exp. Biol. *16:* 389–390 (1957).

173 LASSITTER, J. W.; MORTON, J. D., and MILLER, W. J.: Influence of manganese on skeletal development in the sheep and rat; in MILLS Proc. 1st Int. Symp. Trace Element Metab. Anim., Aberdeen 1969, pp. 130–132 (Livingstone, Edinburgh 1970).

174 LEACH, R. M., jr.; MUENSTER, A.-M., and WIEN, E. M.: Studies on the role of manganese in bone formation. II. Effect on chondroitin sulphate synthesis in chick epiphyseal cartilage. Archs Biochem. Biophys. *133:* 22–28 (1969).

175 LEACH, R. M., jr. and NORRIS, L. C.: Studies on factors affecting the response of chicks to molybdenum (abstr.) Poultry Sci. *36:* 1136 (1957).

176 LEE, C.-C. AND WOLTERINK, L. F.: Blood and tissue partition of cobalt 60 in dogs. Am. J. Physiol. *183:* 173–177 (1955).

177 LEE, C.-C. and WOLTERINK, L. F.: Metabolism of cobalt 60 in chickens. Poultry Sci. *34:* 764–776 (1955).

178 LEE, G. R.; CARTWRIGHT, G. E., and WINTROBE, M. M.: Heme biosynthesis in copper deficient swine. Proc. Soc. exp. Biol. Med. *127:* 977–981 (1968).

179 LEE, H. J. and KUCHEL, R. E.: The aetiology of *Phalaris* staggers in sheep. I. Preliminary observations on the preventative role of cobalt. Austr. J. agric. Res. *4:* 88–99 (1953).

180 LEHMANN, B. H.; HAUSEN, J. D. L., and WARREN, P. J.: The distribution of copper, zinc and manganese in various regions of the brain and in other tissues of children with protein-calorie malnutrition. Br. J. Nutr. *26:* 197–202 (1971).

181 LEVINE, E. P.: Occurrence of titanium, vanadium, chromium and sulphuric acid in the ascidian, *Eudistoma ritteri.* Science *133:* 1352–1353 (1961).

182 LEVINE, H.; REMINGTON, R. E., and KOLNITZ, H. VON: Studies on the relation of diet to goiter. II. The iodine requirement of the rat. J. Nutr. *6:* 347–354 (1933).
183 LEWIS, P. K.; HOEKSTRA, W. G., and GRUMMER, R. H.: Restricted calcium feeding versus zinc supplementation for the control of parakeratosis in swine. J. Anim. Sci. *16:* 578–588 (1957).
184 LIEBERMAN, I. and OVE, P.: Deoxyribonucleic acid synthesis and its inhibition in mammalian cells cultured from the animal. J. biol. Chem. *237:* 1634–1642 (1962).
185 LIEBERMAN, I.; ABRAMS, R.; HUNT, N., and OVE, P.: Levels of enzyme activity and deoxyribonucleic acid synthesis in mammalian cells cultured from the animal. J. biol. Chem. *238:* 3955–3962 (1963).
186 LINDBERG, P.: Selenium determination in plant and animal material, and in water. Acta vet. scand. *9:* suppl. 23 (1968).
187 LONGSTAFF, M. and HILL, R.: The matrix and uronic acid contents of thin and thick shells produced by pullets given diets of varying manganese and phosphorus content; in MILLS Proc. 1st Int. Symp. Trace Element Metab. Anim., Aberdeen 1969, pp. 137–139 (Livingstone, Edinburgh, 1970).
188 LONGSTAFF, M. and HILL, R.: Hexosamine and uronic acid contents of the matrix of shells from pullets fed on diets of different manganese content. Br. Poultry Sci. *13:* 377–385 (1972).
189 LUECKE, R. W.; OLMAN, M. E., and BALTZER, B. V.: Zinc deficiency in the rat. Effect on serum and intestinal alkaline phosphatase activities. J. Nutr. *94:* 344–350 (1968).
190 LUOMA, H.; MEURMAN, J. H.; HELMINEN, S. K. J.; KOSKINEN, K., and RANTA, H.: Modification of dental caries and calculus in rats by fluoride and bicarbonate-phosphate-fluoride additions to dietary sugar. Archs oral Biol. *17:* 821–828 (1972).
191 MAASS, A. R.; MICHAUD, L.; SPECTOR, H.; ELFEHJEM, C. A., and HART, E. B.: The relationship of copper to hematopoeisis in experimental hemorrhagic animals. Am. J. Physiol. *141:* 322–328 (1944).
192 MAGEE, A. C. and MATRONE, G.: Studies on growth, copper metabolism and iron metabolism of rats fed high levels of zinc. J. Nutr. *72:* 233–242 (1960).
193 MAKLID, N. I.: Biological effects of zinc on ducklings. Korma Kormlenie Sel'skokhoz. Zhivotn. *17:* 115–119 (1969); cited in Chem. Abstr. 75–60573 (1971).
194 MALAN, A. I.; TOIT, P. J., DU, and GROENEWALD, J. W.: Iodine in the nutrition of sheep. Final report. Onderstepoort. J. Vet. Sci. anim. Ind. *14:* 329–334 (1940).
195 MARCILESE, N. A.; HARMS, R. H.; VALSECCHI, R. M., and ARRINGTON, L. R.: Iodine uptake by ova of laying hens given excess iodine and effect upon ova development. J. Nutr. *94:* 117–120 (1968).
196 MARSTON, H. R.: Problems associated with 'Coast disease' in South Australia. J. Council Sci. ind. Res., Aust. *8:* 111–116 (1935).
197 MARSTON, H. R.; LEE, H. J., and MCDONALD, H. W.: Cobalt and copper in the nutrition of sheep. J. agric. Sci. *38:* 216–221 (1948).
198 MARTIN, G. R.; PINNELL, S. R.; SIEGEL, R. C., and GOLDSTEIN, E. R.: Lysyl oxidase; in BALASZ The enzyme step in collagen and elastin crosslinking. Chem. Mol. Biol. Intercell Matrix Advance Study Inst. vol. 1, pp. 405–410 (Academic Press, London 1969).
199 MATRONE, G.: Interrelationships of iron and copper in the nutrition and metabolism of animals. Fed. Proc. Fed. Am. Socs exp. Biol. *19:* 659–665 (1960).

200 Matrone, G.; Hartman, R. H., and Clawson, A. J.: Studies of a manganese-iron antagonism in the nutrition of rabbits and baby pigs. J. Nutr. *67:* 309–317 (1959).
201 McClain, P. E.; Wiley, E. R.; Beecher, G. R.; Anthony, W. L., and Hsu, J. M.: Influence of zinc deficiency on synthesis and cross-linking of rat skin collagen. Biochim. biophys. Acta *304:* 457–465 (1973).
202 McCoy, K. E. M. and Weswig, P. H.: Some selenium responses in the rat not related to vitamin E. J. Nutr. *98:* 383–389 (1967).
203 McLean, J. W.; Thomson, G. G., and Claxton, J. H.: Growth responses to selenium in lambs. Nature, Lond. *184:* 251–252 (1959).
204 McQuitty, J. T.; DeWys, W. D.; Monaco, L.; Strain, W. H.; Rob, C. G.; Apgar, J., and Pories, W. J.: Inhibition of tumour growth by dietary zinc deficiency. Cancer Res. *30:* 1387–1390 (1970).
205 Medici, J. C. and Taylor, M. W.: Mineral requirements of the confused flour beetle, *Tribolium confusum* (Duval). J. Nutr. *88:* 181–186 (1966).
206 Medici, J. C. and Taylor, M. W.: Interrelationships among copper, zinc and cadmium in the diet of the confused flour beetle. J. Nutr. *93:* 307–309 (1967).
207 Mertz, W.: Chromium occurrence and function in biological studies. Physiol. Rev. *49:* 164–231 (1969).
208 Mertz, W. and Roginski, E. E.: Chromium metabolism. The glucose tolerance factor; in Mertz and Cornatzer Newer trace elements in nutrition, pp. 123–153 (Marcel Dekker, New York 1971).
209 Mertz, W.; Roginski, E. E., and Riba, R.: Biological activity and fate of trace quantities of intravenous chromium (III) in the rat. Am. J. Physiol. *20:* 489–494 (1965).
210 Messer, H. H.; Armstrong, W. D., and Singer, L.: Fertility impairment in mice on a low-fluoride diet. Science *177:* 893–894 (1972).
211 Messer, H. H.; Armstrong, W. D., and Singer, L.: Influence of fluoride intake on reproduction in mice. J. Nutr. *103:* 1319–1326 (1973).
212 Mildvan, A. S.; Scrutton, M. C., and Utter, M. F.: Pyruvate carboxylase VII. A possible role for tightly-bound manganese. J. biol. Chem. *241:* 3488–3506 (1966).
213 Millar, M. J.; Fisher, M. I.; Elcoate, P. V., and Mawson, C. A.: Effects of dietary zinc deficiency on the reproductive cycle of male rats. Can. J. Biochem. Physiol. *36:* 557–569 (1958).
214 Miller, E. R.; Luecke, R. W.; Ullrey, D. E.; Baltzer, B. V.; Bradley, B. L., and Hoefer, J. A.: Biochemical, skeletal and allometric changes due to zinc deficiency in the baby pig. J. Nutr. *95:* 278–286 (1968).
215 Miller, J. K. and Miller, W. J.: Experimental zinc deficiency and recovery of calves. J. Nutr. *76:* 467–474 (1972).
216 Miller, W. J.; Blackmon, D. M.; Gentry, R. P.; Powell, G. W., and Perkins, H. E.: Influence of zinc deficiency on zinc and dry matter content of ruminant tissues and excretion of zinc. J. Dairy Sci. *49:* 1446–1453 (1966).
217 Miller, W. J.; Morton, J. D.; Pitts, W. J., and Clifton, C. M.: Effect of zinc deficiency and restricted feeding on wound healing in the bovine. Proc. Soc. exp. Biol. Med. *118:* 427–430 (1965).
218 Miller, W. J.; Pitts, W. J.; Clifton, C. M., and Schmittle, S. C.: Experimentally-produced zinc deficiency in the goat. J. Dairy Sci. *47:* 556–559 (1964).

219 MILLS, C. F.: Copper complexes in grassland herbage. Biochem. J. *57:* 603–610 (1954).

220 MILLS, C. F.: Availability of copper in freeze-dried herbage and distribution in chicks and rats. Br. J. Nutr. *9:* 398–408 (1955).

221 MILLS, C. F.: Studies of the copper compounds in aqueous extracts of herbage. Biochem. J. *63:* 187–190 (1956).

222 MILLS, C. F.: The dietary availability of copper in the form of naturally occurring organic complexes. Biochem. J. *63:* 190–193 (1956).

223 MILLS, C. F. and MURRAY, G.: The preparation of a semisynthetic diet low in copper for copper deficiency studies with the rat. J. Sci. Food Agric. *11:* 547–552 (1960).

224 MILLS, C. F.; DALGARNO, A. C.; WILLIAMS, R. B., and QUARTERMAN, J.: Zinc deficiency and the zinc requirement of calves and lambs. Br. J. Nutr. *21:* 751–768 (1967).

225 MILLS, C. F. and FELL, B. F.: Demyelination in lambs born to ewes maintained on high intakes of sulphate and molybdate. Nature, Lond. *185:* 20–22 (1960).

226 MILLS, C. F.; QUARTERMAN, J.; WILLIAMS, R. B.; DALGARNO, A. C., and PANIČ, B.: The effects of zinc deficiency on pancreatic carboxypeptidase activity and protein digestion and absorption in the rat. Biochem. J. *102:* 712–718 (1967).

227 MILLS, C. F. and WILLIAMS, R. B.: Copper concentration and cytochrome oxidase and ribonuclease activities in the brains of copper-deficient lambs. Biochem. J. *85:* 629–632 (1962).

228 MOSKALYUK, L. I.: Role of molybdenum in the etiology of endemic goiter in the Chernovsky region; in PEIVE, Biol. Rol. Molibdena Sb. Tr. Simp., pp. 200–207 (Nauka, Moscow 1968); cited in Chem. Abstr. *78:* 3048 (1973).

229 MOXON, A. L. and RHIAN, M.: Selenium poisoning. Physiol. Rev. *23:* 305–337 (1943).

230 MUTH, O. H.; OLDFIELD, J. E.; REMMERT, L. F., and SCHUBERT, J. R.: Effects of selenium and vitamin E on white-muscle disease. Science *128:* 1090 (1958).

231 MUTH, O. H.; WESWIG, P. H.; WHANGER, P. D., and OLDFIELD, J. E.: Effect of feeding selenium-deficient ration to the sub-human primate *(Saimiri sciureus)*. Am. J. vet. Res. *32:* 1603–1605 (1971).

232 MUTH, O. H.; WHANGER, P. D.; WESWIG, P. H., and OLDFIELD, J. E.: Occurrence of myopathy in lambs of ewes fed added arsenic in a selenium-deficient ration. Am. J. vet. Res. *32:* 1621–1623 (1971).

233 NATION, J. L. and ROBINSON, F. A.: Concentration of some major and trace elements in honeybees, royal jelly and pollens determined by atomic absorption spectrophotometry. J. Apicult. Res. *10:* 35–43 (1971).

234 NAUSS, K. M.: Two forms of dipeptidase in cod muscle. Hoppe-Seyler's Z. physiol. Chem. *352:* 665–673 (1971).

235 NEATHERY, M. W.; MILLER, W. J.; BLACKMON, D. M.; PATE, F. M., and GENTRY, R. P.: Effects of long-term zinc deficiency on feed utilisation, reproductive characteristics and hair growth in the sexually mature goat. J. Dairy Sci. *56:* 98–105 (1973).

236 NEEDHAM, A. E.: Nitrogen metabolism in Annelida; in CAMPBELL Comparative biochemistry of nitrogen metabolism, vol. 1, pp. 207–297 (Academic Press, London 1970).

237 NEILSEN, F. H.: Studies on the essentiality of nickel; in MERTZ and CORNATZER Newer trace elements in nutrition, pp. 215–253 (Marcel Dekker, New York 1971).
238 NESHEIM, M. C. and SCOTT, M. L.: Studies on the nutritive effects of selenium for chicks. J. Nutr. *65:* 601–618 (1958).
239 O'DELL, B. L.; HARDWICK, B. C.; REYNOLDS, G., and SAVAGE, J. E.: Connective tissue defect in the chick resulting from copper deficiency. Proc. Soc. exp. Biol. Med. *108:* 402–405 (1961).
240 O'DELL, B. L.; MORGAN, R. I., and REGAN, W. O.: Interaction of dietary fluoride and magnesium in the guinea-pig. J. Nutr. *103:* 841–850 (1973).
241 O'DELL, B. L.; NEWBERNE, P. M., and SAVAGE, J. E.: Significance of dietary zinc for the growing chicken. J. Nutr. *65:* 503–518 (1958).
242 ORENT, E. R. and MCCOLLUM, E. V.: Effects of deprivation of manganese in the rat. J. biol. Chem. *92:* 651–678 (1931).
243 ORGEBIN-CRIST, M.-C.; FREEMAN, M., and BARNEY, G. H.: Sperm formation in zinc-deficient rats. Ann. Biol. anim. *11:* 547–558 (1971).
244 OTT, E. A.; SMITH, W. H., and HARRINGTON, R. B.: Zinc toxicity in ruminants. II. Effects of high levels of dietary zinc on gains, feed consumption and feed efficiency of beef cattle. J. Anim. Sci. *25:* 419–423 (1966).
245 OTT, E. A.; SMITH, W. H.; HARRINGTON, R. B., and BEESON, W. M.: Zinc toxicity in ruminants. I. Effect of high levels of dietary zinc on gains, feed consumption and feed efficiency of lambs. J. Anim. Sci. *25:* 414–418 (1966).
246 OTT, E. A.: SMITH, W. H.; HARRINGTON, R. B.; STOB, M.; PARKER, H. E., and BEESON, W. M.: Zinc toxicity in ruminants. III. Physiological changes in tissues and alterations in rumen metabolism in lambs. J. Anim. Sci. *25:* 424–431 (1966).
247 PARISI, A. F. and VALLEE, B. F.: Zinc metalloenzymes. Characteristics and significance in biology and medicine. Am. J. clin. Nutr. *22:* 1222–1239 (1969).
248 PATTERSON, E. L.; MILSTREY, R., and STOKSTAD, E. L. R.: Effect of selenium in preventing exudative diathesis in chicks. Proc. Soc. exp. Biol. Med. *128:* 970–972 (1968).
249 PEKAREK, R. S. and BEISEL, W. R.: Effect of endotoxin on serum zinc concentration in the rat. Appl. Microbiol. *18:* 482–484 (1969).
250 PENSACK, J. M. and KLUSSENDORF, R. C.: Poultry Nutr. Conf., Atlantic City 1956.
251 PERDOMO, J. T.; HARMS, R. H., and ARRINGTON, L. R.: Iodine toxicity in poultry. Proc. Soc. exp. Biol. Med. *122:* 758–760 (1966).
252 PERFILEV, G. D.: in MARKOW Dynamics of copper, zinc and manganese during early stages of ontogenesis in sturgeon. Vop. Mirfol. Ekol. Parazitol, Zhivotn. Volgograd 1972, pp. 72–79; cited in Chem. Abstr. *79:* 16103 (1973).
253 PERRY, H. M., jr. and PERRY, E. F.: Normal concentrations of some trace metals in human urine. Changes produced by ethylenediamine-tetra-acetate. J. clin. Invest. *38:* 1452–1463 (1959).
254 PETERS, H. A.: Chelation therapy in acute, chronic and mixed porphyria; in SEVEN and JOHNSON Metal-binding in medicine, pp. 190–199 (J. B. Lippincott, Philadelphia 1960).
255 PFEIFFER, C. C.: Blood histamine, basophil counts and trace elements in the schizophrenias. Contrib. Neurosci. Psychopharmacol., Mol. Syst. Int. Symp., Saint-Laurent 1970, pp. 73–76 (Presses Univ., Montreal 1970).

256 Plumlee, M. P.; Thrasher, D. M.; Beeson, W. N.; Andrews, F. N., and Parker, H. E.: Effects of a manganese deficiency on the growth and development of swine (abstr.) J. Anim. Sci. *13:* 996 (1954).

257 Plumlee, M. P.; Thrasher, D. M.; Beeson, W. N.; Andrews, F. N., and Parker, H. E.: Effects of a manganese deficiency upon the growth, development and reproduction of swine. J. Anim. Sci. *15:* 352–367 (1956).

258 Pond, W. G. and Jones, J. R.: Effect of level of zinc in high calcium diets on pigs from weaning through one reproductive cycle and on subsequent growth of their offspring. J. Anim. Sci. *23:* 1057–1060 (1964).

259 Pories, W. J. and Strain, W. H.: Zinc and wound healing; in Prasad, Zinc metabolism, pp. 378–394 (Ch. C. Thomas, Springfield 1966).

260 Pories, W. J.; Henzel, J. H.; Rob, C. G., and Strain, W. H.: Acceleration of wound healing in man with zinc sulphate given by mouth. Lancet *i:* 121–124 (1967).

261 Poswillo, D. E. and Cohen, B.: Inhibition of carcinogenesis by dietary zinc. Nature, Lond. *231:* 447–448 (1971).

262 Powanda, M. C.; Cockerell, G. L., and Pekarek, R. S.: Amino acid and zinc movement in relation to protein synthesis early in inflammation. Am. J. Physiol. *225:* 399–401 (1973).

263 Prasad, A. S.: Metabolism of zinc and its deficiency in human subjects; in Prasad, Zinc metabolism, pp. 250–303 (Ch. C. Thomas, Springfield 1966).

264 Prasad, A. S.; Miale, A., jr.; Farid, Z.; Sandstead, H. H., and Schulert, A. R.: Zinc metabolism in patients with the syndrome of iron deficiency anemia, hepatosplenomegaly, dwarfism and hypogonadism. J. Lab. clin. Med. *61:* 537–549 (1963).

265 Price, C. A.: Control of processes sensitive to zinc in plants and microorganisms; in Prasad, Zinc metabolism, pp. 69–89 (Ch. C. Thomas, Springfield 1966).

266 Proctor, J. F.; Hogue, D. E., and Warner, R. G.: Selenium, vitamin E and linseed oil meal as preventatives of muscular dystrophy in lambs (abstr.) J. Anim. Sci. *17:* 1183–1184 (1958).

267 Quarterman, J.: The effect of zinc on the uptake of glucose by adipose tissue. Biochim. biophys. Acta *177:* 644–646 (1969).

268 Quarterman, J. and Florence, E.: Observations on glucose tolerance and plasma free fatty acids and insulin in the zinc deficient rat. Br. J. Nutr. *28:* 75–79 (1972).

269 Quarterman, J.; Humphries, W. R.; Morrison, J., and Jackson, F. A.: The effect of zinc deficiency on intestinal and salivary mucins. Biochem. Soc. Transactions *1:* 101 (1973).

270 Reid, B. L.; Kurnick, A. A.; Svacha, R. L., and Couch, J. R.: The effect of molybdenum on chick and poult growth. Proc. Soc. exp. Biol. Med. *93:* 245–248 (1956).

271 Roginski, E. E. and Mertz, W.: Dietary chromium and amino acid incorporation in rats on a low-protein ration (abstr.) Fed. Proc. Fed. Am. Socs exp. Biol. *26:* 301 (1967).

272 Roginski, E. E. and Mertz, W.: An eye lesion in rats fed low chromium diets. J. Nutr. *93:* 249–251 (1967).

273 Roginski, E. E. and Mertz, W.: Effects of chromium (III) supplementation on glucose and amino acid metabolism in rats fed a low-protein diet. J. Nutr. *97:* 525–530 (1969).

274 ROJAS, M. A.; DYER, I. A., and CASSATT, W. A.: Manganese deficiency in the bovine. J. Anim. Sci. *24:* 664–667 (1965).

275 ROSENFELD, I. and BEATH, O. A.: Congenital malformations of eyes of sheep. J. agric. Res. *75:* 93–103 (1947).

276 ROSENFELD, I. and BEATH, O. A.: Effect of selenium on reproduction in rats. Proc. Soc. exp. Biol. Med. *87:* 295–297 (1954).

277 ROTHERY, P.; BELL, J. M., and SPINKS, J. W. T.: Cobalt and vitamin B_{12} in sheep. I. Distribution of radio cobalt in tissue and ingesta. J. Nutr. *49:* 173–181 (1953).

278 ROTRUCK, J. T.; POPE, A. L.; GANTHER, H. E., and HOEKSTRA, W. G.: Prevention of oxidative damage to rat erythrocytes by dietary selenium. J. Nutr. *102:* 689–696 (1972).

279 ROTRUCK, J. T.; POPE, A. L.; GANTHER, H. E.; SWANSON, A. B.; HAFEMAN, D. G., and HOEKSTRA, W. G.: Selenium. Biochemical role as a component of glutathione peroxidase. Science *179:* 588–590 (1973).

280 RUCKER, R. B. and GOTTLIECH-RIEMANN, W.: Isolation and properties of soluble elastin from copper-deficient chicks. J. Nutr. *102:* 563–590 (1972).

281 RUCKER, R. B.; GOTTLIECH-RIEMANN, W.; HOBE, J., and DEVERS, K.: Valine incorporation into elastin-rich fractions from chick aorta. Biochim. biophys. Acta *279:* 213–216 (1972).

282 RUCKER, R. B.; PARKER, H. E., and ROGLER, J. C.: Effect of copper deficiency on chick bone collagen and selected bone enzymes. J. Nutr. *98:* 57–63 (1969).

283 RUZICKA, J. A.; MRKLAS, L., and ROKYTOVA, K.: Influence of water intake on the degree of fluorosis and on the incorporation of fluoride into bones and incisor teeth of mice. Caries. Res. *7:* 166–172 (1973).

284 SALISBURY, R. M.; EDMONSON, J.; POOLE, W. S. H.; BOBBY, F. C., and BERNIE, H.: Exudative diathesis and white-muscle disease of poultry in New Zealand. Proc. 12th World's Poultry Congr., Sydney 1962, pp. 379–384.

285 SAMILKIN, N. S. and VOROB'EV, V. I.: Effect of some trace elements on the growth of white Amur roe. Ryb. Khoz., Moscow *8:* 29–31 (1972); cited in Chem. Abstr. *77:* 163293 (1972).

286 SANDSTEAD, H. H.; GILLESPIE, D. D., and BRADY, R. N.: Zinc deficiency. Effect on brain of suckling rat. Pediat. Res. *6:* 119–125 (1972).

287 SANDSTEAD, H. H. and SHEPHARD, G. H.: The effect of zinc deficiency on the tensile strength of healing surgical incisions in the integument of the rat. Proc. Soc. exp. Biol. Med. *128:* 687–689 (1968).

288 SARASWAT, R. C. and ARORA, S. P.: Effect of dietary zinc on the vitamin A and alkaline phosphatase in blood sera of lambs. Indian J. Anim. Sci. *42:* 358–362 (1972).

289 SAVINA, P. N.: Importance of manganese in the formation of endemic goitre. Tr. Tomsk Nauch. – Issled. Inst. Vaktsin Syvorotok. Tomsk. med. Inst. *20:* 183–185 (1969); cited in Chem. Abstr. *75:* 96596 (1971).

290 SCHAIBLE, P. J. and BANDEMER, S. L.: The effect of mineral supplements on the availability of manganese. Poultry Sci. *21:* 8–14 (1942).

291 SCHROEDER, H. A.: Serum cholesterol and glucose levels in rats fed refined and less refined sugars and chromium. J. Nutr. *97:* 237–242 (1969).

292 SCHROEDER, H. A. and BALASSA, J. J.: Influence of chromium, cadmium and lead on rat aortic lipids and circulating cholesterol. Am. J. Physiol. *209:* 433–437 (1965).

293 SCHROEDER, H. A. and BALASSA, J. J.: Arsenic, germanium, tin and vanadium in mice. Effects on growth, survival and tissue levels. J. Nutr. *92:* 245–252 (1967).

294 SCHROEDER, H. A.; BALASSA, J. J., and TIPTON, I. H.: Abnormal trace metals in man. Vanadium. J. chron. Dis. *16:* 1047–1071 (1963).

295 SCHROEDER, H. A.; BALASSA, J. J., and TIPTON, I. H.: Abnormal trace metals in man. Tin. J. chron. Dis. *17:* 483–502 (1964).

296 SCHROEDER, H. A.; BALASSA, J. J., and TIPTON, I. H.: Essential trace metals in man. Manganese. J. chron. Dis. *19:* 545–571 (1966).

297 SCHROEDER, H. A.; KANISAWA, M.; FROST, D. V., and MITCHENER, M.: Germanium, tin and arsenic in rats. Effect on growth, survival, pathologic lesions and life span. J. Nutr. *96:* 37–45 (1968).

298 SCHROEDER, H. A.; NASON, A. P.; TIPTON, I. H., and BALASSA, J. J.: Essential trace metals in man. Copper. J. chron. Dis. *19:* 1007–1034 (1966).

299 SCHROEDER, H. A.; NASON, A. P.; TIPTON, I. H., and BALASSA, J. J.: Essential trace metals in man. Zinc. Relation to environmental cadmium. J. chron. Dis. *20:* 179–210 (1967).

300 SCHROEDER, H. A.; VINTON, W. H., jr., and BALASSA, J. J.: Effect of chromium, cadmium and lead on serum cholesterol of rats. Proc. Soc. exp. Biol. Med. *109:* 859–860 (1962).

301 SCHUBERT, W. K. and LAHEY, M. E.: Copper and protein depletion complicating hypoferric anemia of infancy. Pediatrics *24:* 710–733 (1959).

302 SCHWARTZ, K.: Tin as an essential growth factor for rats; in MERTZ and CORNATZER Newer trace elements in nutrition, pp. 313–326 (Marcel Dekker, New York 1971).

303 SCHWARTZ, K.: Growth-promoting effects of silicon in rats. Nature, Lond. *239:* 333–334 (1972).

304 SCHWARTZ, K.: A bound form of silicon in glycosaminoglycans and polyuronides. Proc. natn. Acad. Sci USA *70:* 1608–1612 (1973).

305 SCHWARTZ, K.; BIERI, J. G.; BRIGGS, G. M., and SCOTT, M. L.: Prevention of exudative diathesis in chicks by factor 3 and selenium. Proc. Soc. exp. Biol. Med. *95:* 621–625 (1957).

306 SCHWARTZ, K. and FOLTZ, C. M.: Selenium as an integral part of factor 3 against dietary necrotic liver degeneration. J. Am. chem. Soc. *79:* 3292–3293 (1957).

307 SCHWARTZ, K. and MERTZ, W.: A glucose tolerance factor and its differentiation from factor 3. Archs Biochem. Biophys. *72:* 515–518 (1957).

308 SCHWARTZ, K. and MERTZ, W.: Chromium (III) and the glucose tolerance factor. Archs Biochem. Biophys. *85:* 292–295 (1959).

309 SCHWARTZ, K. and MILNE, D. B.: Growth effect of vanadium in the rat. Science *174:* 426–428 (1971).

310 SCOTT, M. L.; HILL, F. W.; NORRIS, L. C.; DOBSON, D. C., and NELSON, T.: Studies on vitamin E in poultry nutrition. J. Nutr. *56:* 387–402 (1957).

311 SCOTT, M. L.; HOLM, E. R., and REYNOLDS, R. E.: Studies on the niacin, riboflavine, choline, manganese and zinc requirements of young ring-necked pheasants for growth, feathering and prevention of leg disorders. Poultry Sci. *38:* 1344–1350 (1959).

312 SCOTT, M. L.; OLSON, G.; KROOK, L., and BROWN, W. R.: Selenium-responsive myopathies of myocardium and of smooth muscle in the young poult. J. Nutr. *91:* 573–583 (1967).

313 SCRUTTON, M. C.; GRIMINGER, P., and WALLACE, J. C.: Pyruvate carboxylase. Bo metal content of the vertebrate liver enzyme as a function of diet and species. J. b Chem. *247:* 3305–3313 (1972).

314 SEELIG, M. S.: Relations of copper and molybdenum to iron metabolism. Am. J. C Nutr. *25:* 1022–1037 (1972).

315 SHAND, A. and LEWIS, G.: Chronic copper poisoning in young calves. Vet. Rec. *69:* 618–620 (1957).

316 SHARPLESS, G. R. and METZGER, M.: Arsenic and goiter. J. Nutr. *21:* 341–346 (19

317 SHILS, M. E. and McCOLLUM, E. V.: Further studies on the symptoms of mangan deficiency in the rat and mouse. J. Nutr. *26:* 1–19 (1943).

318 SHRADER, R. E. and EVERSON, G. J.: Anomalous development of otoliths associa with postural defects in manganese-deficient guinea-pigs. J. Nutr. *91:* 453– (1967).

319 SIEGEL, R. C.; RINNELL, S. R., and MARTIN, G. R.: Cross-linking of collagen a elastin. Properties of lysyl oxidase. Biochemistry *9:* 4486–4492 (1970).

320 SINGH, P.: Bibliography of artificial diets for insects and mites. Bull. 209, New Z land Dept. Sci. Ind. Res. (1972).

321 SLATER, J. P.; MILDVAN, A. S., and LOEB, L. A.: Zinc in DNA polymerases. B chem. biophys. Res. Commun. *44:* 37–43 (1971).

322 SMITH, J.; McDANIEL, E. G.; McBEAN, L. D.; DOFT, F. S., and HALSTEAD, J. A Effect of microorganisms upon zinc metabolism, using germ-free and conventio rats. J. Nutr. *102:* 711–719 (1972).

323 SMITH, J. C. and HACKLEY, B.: Distribution and excretion of nickel-63 administer intravenously to rats. J. Nutr. *95:* 541–546 (1968).

324 SMITH, J. C., jr.; McDANIEL, E. G.; FAN, F. F., and HALSTEAD, J. A.: Zinc. Tra element essential in vitamin A metabolism. Science *181:* 954–955 (1973).

325 SMITH, S. E. and MEDLICOTT, M.: The blood picture of iron and copper deficien in the rat. Am. J. Physiol. *141:* 354–358 (1944).

326 SMITH, S. E.; MEDLICOTT, M., and ELLIS, G. H.: Manganese deficiency in the rabb Archs Biochem. *4:* 281–289 (1944).

327 SMITH, S. E.; MEDLICOTT, M., and ELLIS, G. H.; The blood pictures of iron a copper deficiency anemias in the rabbit. Am. J. Physiol. *142:* 179-181 (1944).

328 SÖREMÄRK, R.: Vanadium in some biological specimens. J. Nutr. *93:* 183–19 (1967).

329 SPALLHOLZ, J. E.; MARTIN, J. L.; GERLACK, M. H., and HEINZERLING, R. H.: In munologic response of mice fed diets supplemented with selenite selenium. Pro Soc. exp. Biol. Med. *143:* 685–689 (1973).

330 SPENCER, H. and SAMACHSON, J.: Studies of zinc metabolism in man; in MILLS Pro 1st Int. Symp. Trace Element Metab. Anim., Aberdeen 1969, pp. 312–317 (Livingstone, Edinburgh 1970).

331 SPIVEY-FOX, M. P. and HARRISON, B. N.: Use of Japanese quail for the study of zin deficiency. Proc. Soc. exp. Biol. Med. *116:* 256–259 (1964).

332 SPRAY, C. M. and WIDDOWSON, E. M.: Effect of growth and development on th composition of mammals. Br. J. Nutr. *4:* 332–353 (1951).

333 SPRINKER, L. H.; HARR, J. R.; NEWBERNE, P. M.; WHANGER, P. D., and WESWIG

P. H.: Selenium-deficiency lesions in rats fed vitamin E-supplemented rations. Nutr. Rep. Int. *4:* 335–340 (1971).

4 Srirangereddy, G. and Rao, B. S. N.: Effect of fluoride on the skeleton of rats maintained on different levels of calcium in the diet. Indian. J. med. Res. *60:* 481–487 (1972).

5 Srivastava, P. N. and Auclair, J. L.: Improved chemically defined diet for the pea aphid *Acyrthosiphon pisum.* Ann. entomol. Soc. Am. *64:* 474–478 (1971).

6 Starcher, B. C.: Studies on the mechanism of copper absorption in the chick. J. Nutr. *97:* 321–326 (1969).

7 Sturgeon, P. and Brubaker, C.: Copper deficiency in infants. A syndrome characterised by hypocupremia, iron deficiency anemia and hypoprotinemia. Am. J. Dis. Child. *92:* 254–265 (1956).

8 Stephan, J. K. and Hsu, J. M.: Effect of zinc deficiency and wounding on DNA synthesis in rat skin. J. Nutr. *103:* 548–552 (1973).

9 Sullivan, J. F. and Lankford, H.E.: Urinary excretion of zinc in alcoholism and postalcoholic cirrhosis. Am. J. Clin. Nutr. *10:* 153–157 (1962).

0 Sultanova, G. F.: Peculiarities in copper metabolism in prematurely born infants in the first months of life. Pediatrija *10:* 14–18 (1970).

1 Summers, J. D. and Moran, E. T.: Interactions of dietary vanadium, calcium and phosphorus for the growing chicken. Poultry Sci. *51:* 1760–1761 (1972).

2 Suttie, J. W.; Phillips, P. H., and Miller, R. F.: Studies on the effects of dietary sodium fluoride on dairy cows. III. Skeletal and soft tissue fluorine deposition and fluorine toxicosis. J. Nutr. *65:* 293–304 (1958).

3 Suttle, N. F. and Mills, C. F.: Studies of the toxicity of copper to pigs. I. Effects of oral supplements of zinc and iron salts on the development of copper toxicosis. Br. J. Nutr. *20:* 135–148 (1966).

44 Suttle, N. F. and Mills, C. F.: Studies of the toxicity of copper to pigs. II. Effect of protein source and other dietary components on the response to high and moderate intakes of copper. Br. J. Nutr. *20:* 149–161 (1966).

45 Swenerton, H. and Hurley, L. S.: Severe zinc deficiency in male and female rats. J. Nutr. *95:* 8–18 (1968).

46 Swenerton, H.; Shrader, R., and Hurley, L. S.: Zinc-deficient embryos: reduced thymidine incorporation. Science *166:* 1014–1015 (1969).

47 Thomson, G. G. and Lawson, B. M.: Copper and selenium interactions in sheep. N. Z. vet. J. *18:* 79–82 (1970).

48 Tipton, I. H.: The distribution of trace metals in the human body; in Seven and Johnson Metal-binding in medicine, pp. 27–42 (J. B. Lippincott, Philadelphia 1960).

49 Tucker, H. F. and Salmon, W. D.: Parakeratosis or zinc deficiency disease in the pig. Proc. Soc. exp. Biol. Med. *88:* 613–616 (1955).

350 Ullrey, D. E.; Miller, E. R.; Thomson, O. A.; Zutaut, C. L.; Schmidt, D. A.; Ritchie, H. D.; Hoefer, J. A., and Luecke, R. W.: Studies of copper utilisation by the baby pig. (abstr.) J. Anim. Sci. *19:* 1298 (1960).

351 Underwood, E. J.: Trace elements in human and animal nutrition; 3rd ed. (Academic Press, London 1971).

352 Underwood, E. J. and Filmer, J. F.: Determination of the biologically potent element (cobalt) in limonite. Aust. vet. J. *11:* 84–92 (1935).

353 UNDERWOOD, E. J. and SOMERS, M.: Studies of zinc nutrition in sheep. I. The relation of zinc to growth, testicular development and spermatogenesis in young rams. Aust. J. agric. Res. *20:* 889–897 (1969).

354 WACHTEL, L. W.; ELVEHJEM, C. A., and HART, E. B.: Studies on the physiology of manganese in the rat. Am. J. Physiol. *140:* 72–82 (1943).

355 WAHLSTROM, R. C. and OLSON, D. E.: The effect of selenium on reproduction in swine. J. Anim. Sci. *18:* 141–145 (1959).

356 WARKANY, J. and PETERING, H. G.: Malformations of the central nervous system in rats produced by maternal zinc deficiency. Teratology *5:* 319–334 (1972).

357 WARKANY, J. and PETERING, H. G.: Congenital malformations of the brain produced by short zinc deficiencies in rats. Am. J. ment. Defic. *77:* 645–653 (1973).

358 WEBER, C. W. and REID, B. L.: Nickel toxicity in growing chicks. J. Nutr. *95:* 612–616 (1968).

359 WEBER, C. W. and REID, B. L.: Nickel toxicity in young growing mice. J. Anim. Sci. *28:* 620–623 (1969).

360 WEDDELL, J.; STEENBOCK, H., and HART, E. B.: Growth and reproduction upon milk diets. J. Nutr. *4:* 53–65 (1931).

361 WESER, U.; SEEBER, S., and WARNECKE, P.: Reactivity of Zn^{2+} on nuclear DNA and RNA biosynthesis of regenerating rat liver. Biochim. biophys. Acta *179:* 422–428 (1969).

362 WHANGER, P. D.: Effects of dietary nickel on enzyme activities and mineral contents in rats. Toxic. appl. Pharmac. *25:* 323–331 (1973).

363 WIDDOWSON, E. M. and SPRAY, C. M.: Chemical development *in utero*. Archs Dis. Childh. *26:* 205–214 (1951).

364 WILGUS, H. S., jr: The role of manganese and certain other trace elements in the prevention of perosis. J. Nutr. *14:* 155–167 (1937).

365 WILLIAMS, R. B.: Intestinal alkaline phosphalase and inorganic pyrophosphate activities in the zinc-deficient rat. Br. J. Nutr. *27:* 121–130 (1972).

366 WILLIAMS, R. B. and CHESTERS, J. K.: The effects of early zinc deficiency on DNA and protein synthesis in the rat. Br. J. Nutr. *24:* 1053–1059 (1970).

367 WILLIAMS, R. B.; DEMERTZIS, P., and MILLS, C. F.: Effects of dietary zinc concentration on reproduction in the rat. Proc. Nutr. Soc. *32:* 3A (1973).

368 WILLIAMS, R. B. and MILLS, C. F.: The experimental production of zinc deficiency in the rat. Br. J. Nutr. *24:* 989–1003 (1970).

369 WILLIAMS, R. B.; MILLS, C. F.; QUARTERMAN, J., and DALGARNO, A. C.: The effect of zinc deficiency on the *in vivo* incorporation of ^{32}P into rat-liver nucleotides. Biochem. J. *95:* 29P (1965).

370 WILLIAMS, R. B.; MILLS, C. F., and DAVIDSON, R. J. L.: Relationships between zinc deficiency and folic acid status of the rat. Proc. Nutr. Soc. *32:* 2A (1973).

371 WOOD, E. C. and WORDEN, A. N.: Influence of dietary copper concentration on hepatic copper in the duckling and chick. J. Sci. Food Agric. *24:* 167–174 (1973).

372 WRIGHT, P. L. and BELL, M. C.: Comparative metabolism of selenium and tellurium in sheep and swine. Am. J. Physiol. *211:* 6–10 (1966).

373 WU, S. H.; OLDFIELD, J. E.; WHANGER, P. D., and WESWIG, P. H.: Effect of selenium, vitamin E and antioxidants on testicular function in rats. Biol. Reprod. *8:* 625–629 (1973).

374 YARRINGTON, J. T.; WHITEHAIR, C. K., and CORWIN, R. M.: Vitamin E – selenium deficiency and its influence on avian malarial infection in the duck. J. Nutr. *103:* 231–241 (1973).

375 ZEL'STER, M. E. and NIKOV, P. S.: Phagocytic activity and indexes of neutrophil metabolism in iodine deficiency. Gig. Sanit. *37:* 99–101 (1972); cited in Chem. Abstr. *77:* 60395 (1972).

R. B. WILLIAMS, Department of Nutritional Biochemistry, The Rowett Research Institute, Bucksburn, *Aberdeen AB2 9SB* (Scotland)

Comp. Anim. Nutr., vol. 2, pp. 144–178 (Karger, Basel 1977)

Activity Spectrum of Ingested Toxicants

T. D. LUCKEY

Department of Biochemistry, University of Missouri Medical School, Columbia, Mo.

I. General Consideration

A. Nutrition Orientation

The comparative viewpoint of ingested toxicants comes from comparative biochemistry, physiology and nutrition. The unity of all living organisms is exemplified by their relatively constant composition of tissues; excepting hard tissues of support structures, only the most sophisticated chemical analyses could distinguish the tissues of one species from those of any other. This similarity in chemical composition is reflected in common metabolic pathways by which conservative evolution records phylogenetic progress. Success in feeding a single diet to representatives of many phyla [LUCKEY, 1960] illustrates the unity of nutrition from unicellular forms to primates. Diversity in comparative nutrition was hidden in that study because most nutrients were available irrespective of their requirement. Diversity also was noted by the failures as discussed in the introductory chapter. Comparable facets of unity and diversity are seen in the food chains of nature.

Nurture comprises all environmental influences upon the individual. From this philosophic viewpoint nutrition is the chemical component of the environment which affects the individual. This includes material ingested, inhaled, absorbed through the skin, and/or injected via wounds or natural orifices. A toxicant is any chemical which directly or indirectly impairs the normal functioning of cells and tissues. Toxicants can be accepted as components of nutrition because all nutrients are toxic when given in excess, most toxicants are ingested, and many toxicants are stimulatory when administered in very low doses. The known harmful affects of each essential nutrient are documented in the last section of each chapter of this treatise. Definitive data regarding the toxicity and lethality of ingested essential nutrients are not readily available. Although ingested toxicants are the major subject of this review, materials transferred across the gills and skin of aquatic animals or toxins introduced into the trachea of invertebrates are considered also. The dynamic processes through which animals are nurtured (fig. 1) include the processing of ingesta by microbes in the alimentary tract and the reintroduction of finite quantities of their excreta.

VENUGOPAL and LUCKEY [1974] have modified the law of optimal nutritive concentration (fig. 2) attributed to BERTRAND as a dose-response continuum. A new truth emerges: the dose-response relationships which are usually considered separately constitute a trinity. First is the dose-

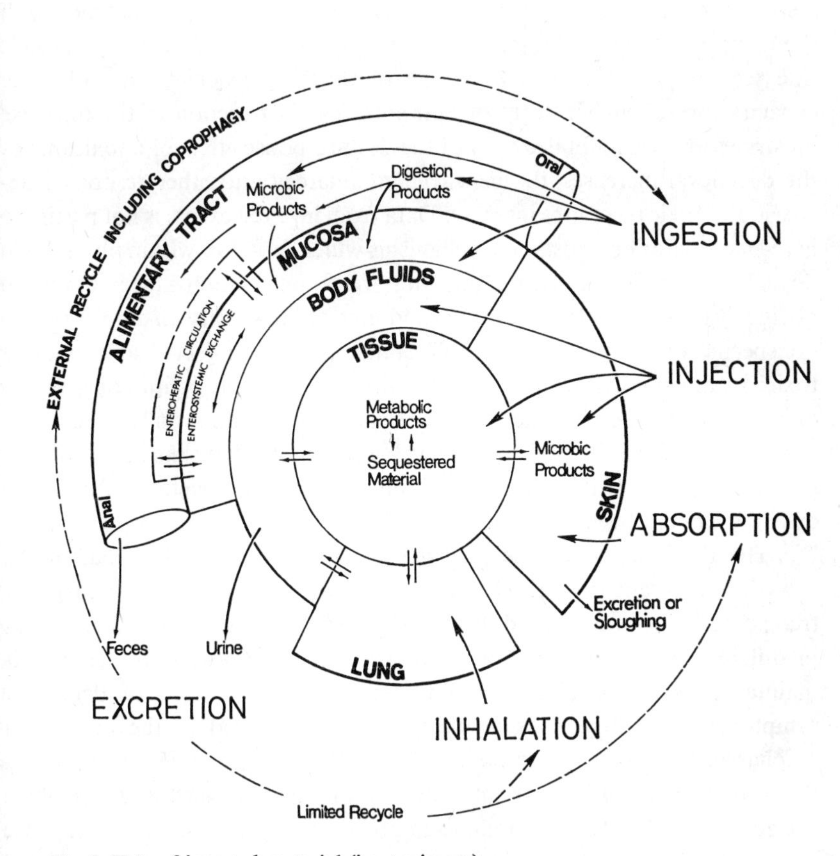

Fig. 1. Fate of ingested material (i.e. toxicants).

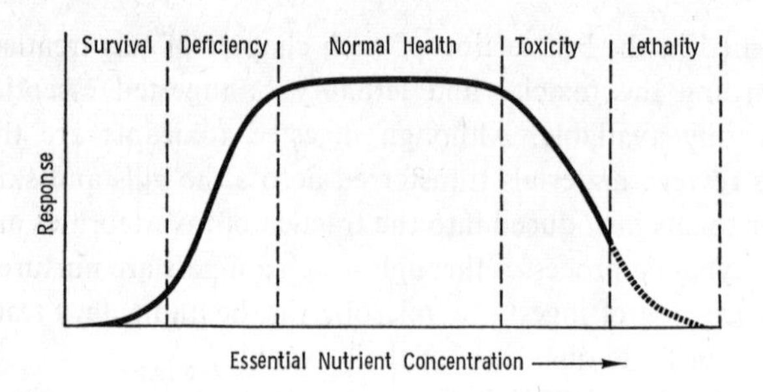

Fig. 2. Optimal nutrient concentration [VENUGOPAL and LUCKEY, 1975].

response of the essential nutrient. Nutritionists are usually concerned with the first part of the response, the essential nature and deficiency symptoms. The second part, the plateau region of the curve, depicts the tolerance capacity toward higher doses of nutrients. The last section of the response illustrates an oversimplified view of the dose-response effect of a toxicant; as the dose level increases, the beneficial stimulatory and other responses decrease and toxic responses increase. Data from injected excess is not pertinent here since major homeostatic mechanisms which function within physiologic doses have been bypassed; the time-dose-response relationships in tissue are obviously different between injected and ingested toxicants. Examples of two nonspecific mechanisms which differentiate between injected and ingested toxicities are cited. Rats fed excessive vitamin C exhibit continuous, severe diarrhea prior to death. Chicks refuse to eat a diet containing a lethal concentration (1% of the diet) of oxytetracycline. Both mechanisms are obviously nonspecific, and neither would function as protective reactions if the same dose per kg had been injected.

The third dose-response relationship to be considered is that of the growth promotant (fig. 3). The contrast of the effect in classic and germfree animals illustrates the different response patterns of the same species in different environments. This response may disappear if the subject is maintained under optimum conditions; this and the lack of deficiency symptoms in healthy stock resulting from withdrawal of the promotant emphasize the lack of essentiality of growth promotants [LUCKEY, 1959]. Paradoxically, essential nutrients can be promotants at dietary levels which are abnormal. The two possibilities imposed upon the classic dose-response relationship are illustrated in figure 4. Specific examples to be presented

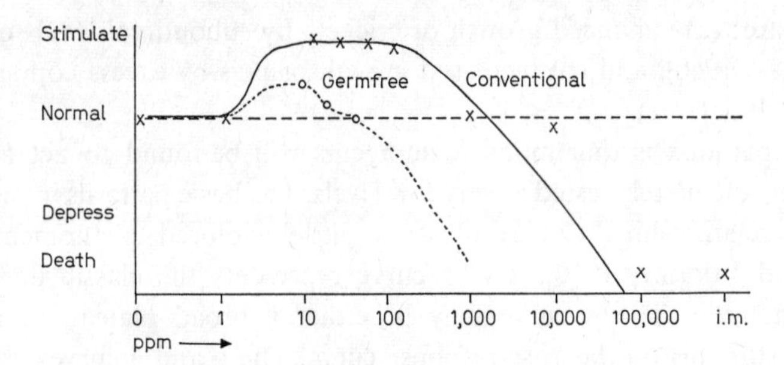

Fig. 3. Effect of oxytetracyclene upon the growth rate of germfree and conventional chicks. From LUCKEY [1958].

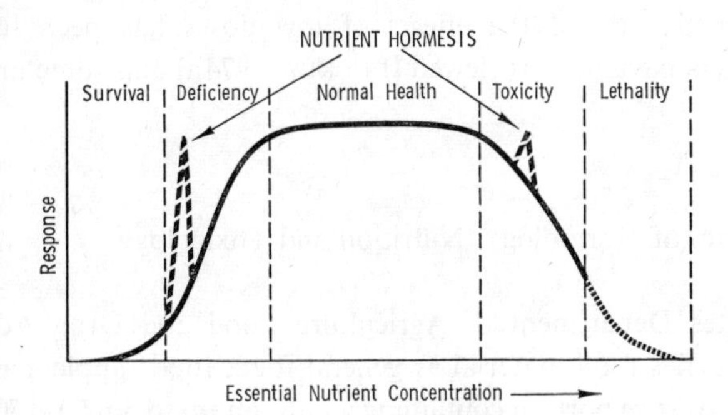

Fig. 4. Hormology of essential nutrients have been noted with Na and Cu.

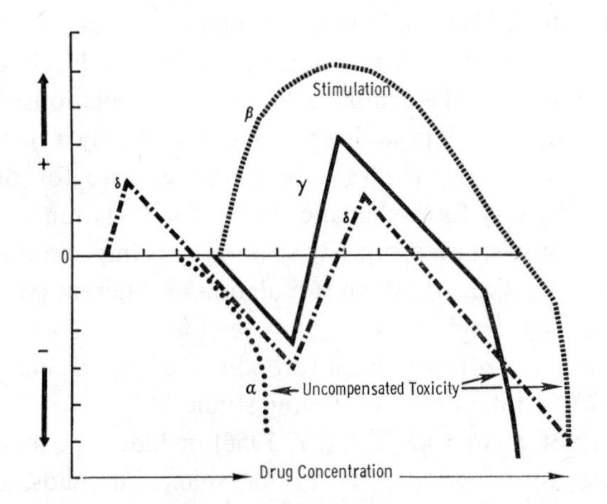

Fig. 5. Different types of dose-response curves on a semi-log basis to elucidate different phenomena of hormology. The ordinate is *response*.

later in detail are: (a) enhanced growth of crickets by suboptimal levels of sodium [LUCKEY, 1960], and (b) increased growth of pigs by excess copper [BARBER *et al.*, 1957].

A second paradox is that many toxic agents will be found to act as stimulants when adequately tested at very low levels. The basic patterns in the dose-response relationships (fig. 5) illustrate little explored phenomena [TOWNSEND and LUCKEY, 1960]. The α curve represents the classic dose response for a toxin. The β curve may be either a broad plateau or a sharp peak of stimulus on the dose-response curve. The γ and δ curves are complex, difficult to interpret, and are noted sporadically; each exhibits depressed functions at doses which are less than stimulatory doses. Specific examples will be noted later. General unfamiliarity with these patterns suggests that objective study of the effects of low doses has been inadequate. Some aspects have been reviewed [LUCKEY, 1974a] and some are presented herein.

B. The Continuum of Hormology, Nutrition and Toxicology

The United States Department of Agriculture Food and Drug Administration now classifies food material as general food, food supplement or drug, based on an average portion containing a nutrient up to 50, 50–150, or over 150% of the recommended daily allowance (RDA). This blending of food into drug depends upon the highest amount of any essential nutrient present in the food relative to RDA; is one eating a food or a drug?

The toxicity of excess nutrients and the stimulation by low levels of toxicants is a valid basis to consider the complete dose-response relationship as a continuum. This does not invalidate studies which consider only a small part of the dose-response curve; it provides a broad perspective for the limited studies and opens the way for systematic study of stimulation as a component of the dose-response relationship. Stimulation is as important as toxicity in both theory and practice. Although stimulation of different parameters is often found with low levels of toxicants over long time feeding, this phenomenon is separate and distinct from recondite toxicity of metals noted by SCHROEDER [1973] in long life term feeding studies.

Hormology, the study of excitation [LUCKEY, 1956] includes the stimulatory effects of low doses of many inorganic and organic compounds, all of which would be toxic at higher dose levels. The stimulatory agent is a *hormetic* and the action has been called *hormesis* [LUCKEY, 1974b]. Thus, the complete biologic activity of most compounds involves stimulatory as well as nutritional and/or toxicologic interpretations. The continuum of the dose-response relationship amalgamates hormology, nutrition and toxicology into one conceptual unity. A compound is not 'toxic' or 'stimulatory'; the action of any compound depends upon the dose of the compound, the species to which it is administered, and the conditions of the organism and its environment. These variables are of paramount importance when a stimulatory response or a recondite toxicity is being investigated. LUCKEY [1975] has reviewed hormology with inorganic compounds; details of these effects may be contrasted with metal ion toxicities reviewed by VENUGOPAL and LUCKEY [1975].

The continuum of the dose-response relationship from hormology through toxicity provides an important concept and a definition of zero equivalent point (ZEP) for the use of regulatory and administrative departments of government. ZEP is that dose level of a compound which is biologically equivalent to the zero concentration of the toxicant for the character being studied. Any agent which gives a stimulation (β curve or equivalent point on γ and δ curves) can provide adequate experimental data to give a ZEP when the curve is drawn between the highest stimulating dose and the lowest inhibitory dose. For example, the ZEP dose of oxytetracycline for classic chicks is 10,000 ppm (fig. 3). For each parameter, or set of parameters with each having preweighed or assigned partial values, the dose level at which the curve crosses the zero response line is experimentally equivalent to a dose level at which the agent is physiologically inactive. The physiologic response or the sum of all metabolic reactions as measured by the parameters studied is the same at zero dose level as at ZEP. Time and accumulative dosage are considerations to be standardized. In nutrition the drug is usually added to food or water for a long period; in pharmacology a single dose is often administered. Both provide acceptable time-dose-response data.

C. Ingested Toxicants

For comparative animal nutrition ingested toxicants may be grouped as: (1) essential nutrients; (2) antimetabolites; (3) food toxicants, and (4) other toxicants. Antimetabolites are retained in a separate group because of their special characteristics; some could be grouped under both food toxicants and other ingested toxicants.

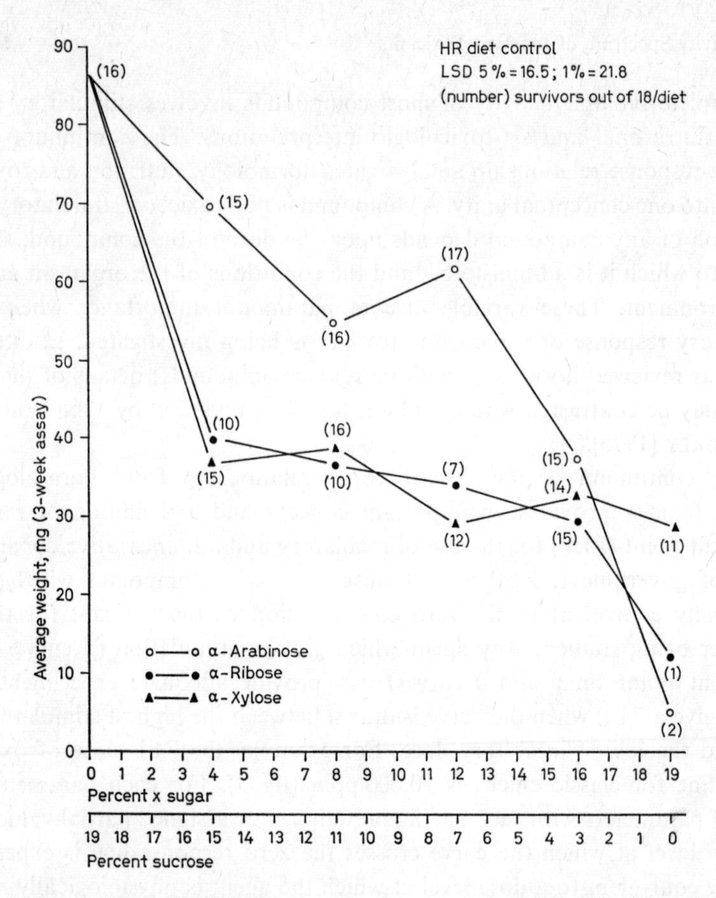

Fig. 6. Effect of exchanging pentoses for sucrose in the diet of crickets. Taken from NEVILLE and LUCKEY [1962].

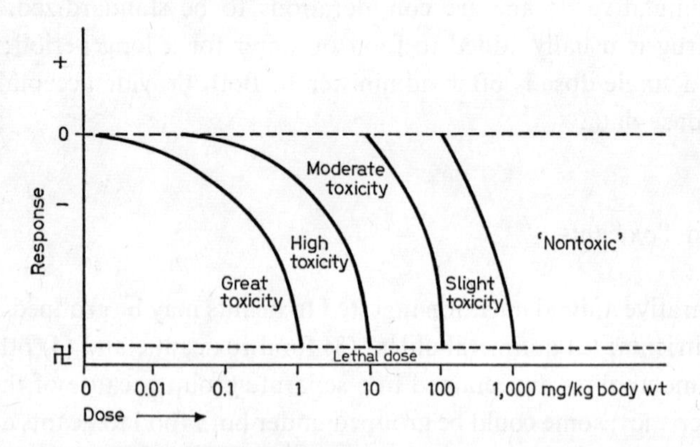

Fig. 7. Classification of toxicants on the basis of lethal doses.

1. Essential Nutrients

Although little experimental data is available from toxicity studies with ingested nutrients, each of the 50 ± essential nutrients is harmful when fed in excess. Even water can become toxic if the usual consumption is quadrupled. Pharmacologic effects and toxicity syndromes observed after parenteral injection may differ dramatically from those noted following excessive feeding of the same nutrient. The inherent toxicity of essential nutrients depends on the ratio of essential to toxic doses or therapeutic to toxic doses which is represented by the extent of the plateau of normal health (fig. 2). Some nutrients, i.e. selenium, have a very narrow plateau; these may be known better as harmful substances than as nutrients. Other nutrients elicit a broad plateau and may never have been fed in sufficiently high quantity to determine the toxicity limits or syndromes; or the organism may have a good homeostasis for that nutrient, e.g. vitamin C in man. The experimental work with vitamin A toxicity in laboratory animals and recent practical problems with vitamin D toxicity in infants illustrates the fact that even nutrients having a broad 'normal health' plateau may show toxicity. The minimum lethal dose of a nutrient is often 1,000 times the recommended daily allowance. A compound which is a nutrient for one species may be harmful to another at physiologic doses; pentoses fed at 5% of the diet depressed the growth of crickets and at 19% they increased the morbidity (fig. 6).

Arbitrary criteria for classifying toxicants on the basis of LD_{50} values (fig. 7) may suggest the relative toxicity of compounds. By this criteria most essential nutrients could be categorized as nontoxic. However, excessive intakes of any material for a prolonged period will develop harmful and/or toxic effects. A recent example is excessive pyridoxine was found to induce a 40% increment in liver mass [COHEN *et al.*, 1973]. Our laboratory confirmed the work of DAFT and associates [MCDANIEL and DAFT, 1954] who noted that rats fed 3% vitamin C developed diarrhea; when the dietary level of vitamin C was increased to 4%, mortality became high.

An excess of one nutrient can inhibit or prevent utilization of another essential nutrient. Imbalance between essential nutrients indirectly may lead to a deficiency of another nutrient. Thus, imbalance causes a deficiency of another nutrient; this is one mechanism of toxicity for some nutrients given in excess. A variety of interactions within the alimentary tract may cause one nutrient to limit the absorption of another, i.e. excess fat or Ca^{++} causes the formation of soaps or insoluble phosphates which cause poor absorption of fat-soluble vitamins and minerals, such as Zn. Essential nutrients may be antagonistic by direct competition for macromolecular sites, i.e. the competition of valine, leucine, and isoleucine or Fe, Mn, Ni, Cu, and Zn for common active or facilitated transport mechanism, or by indirect competition, i.e. the balancing of cell permeability by appropriate ratio of Na^+ or Ca^{++} ions. The observed symptoms constitute a toxicity of the excessive nutrient; normal growth and/or metabolism is depressed.

Chemical antagonisms include heavy metal precipitation, agglutination and chelation reactions and nondissociated complex formation in which the high affinity of the toxicants for active sites of functional macromolecules inactivates enzymes or other essential molecules. The chelation of inositol hexaphosphate (phytic acid) to minerals is one example of the great variety of chemical antagonisms.

2. Antimetabolites

Antimetabolites constitute a special class in which the molecular configuration of the toxicant resembles that of an essential metabolite. These toxicants may compete with the metabolite for a site on a functioning macromolecule, or they may modify a macromolecule by attachment to an allosteric site.

Antimetabolites may interfere with an enzyme or cell receptor because of competitive or noncompetitive antagonism to a specific metabolite. Competitive inhibitors have the greatest similarity of structure and electrical charge distribution; the best known examples are sulfanilamide and *p*-aminobenzoic acid. The similarity allows the inhibitor to react with the receptor (i.e. enzyme, carrier, or membrane) macromolecule:

$$R + I \leftrightharpoons RI,$$

where R = receptor macromolecule and I = inhibitor, with no further development. The natural metabolite-receptor complex can dissociate into products:

$$R + M \leftrightharpoons RM \leftrightharpoons R + P_1 \pm P_2$$

where M = metabolite or substrate and P = product or a transferred metabolite. The equilibrium of each reaction depends upon the affinity of R for either I or M. In a mixture a high affinity RI would demand a relatively large quantity of M to reverse the reaction to form RM. Thus, the degree of inhibition depends upon the receptor affinities for and the relative concentrations of inhibitor and metabolite. If enough metabolite were added, maximum velocity may be obtained in the presence of inhibitor.

Table I. Selected examples of antimetabolites[1]

Metabolite class	Metabolite	Inhibitor	Organism	Reference
Amino acid	methionine	nor-aline	*Proteus morgani*	PORTER and MEYERS [1945]
	valine	α-aminoisobutane sulfonic acid	*Proteus vulgaris*	MCILWAIN [1941]
	cysteine	selenium cystine	rats (tumor)	WEISBERGER and SUHRLAND [1955]
Vitamin	thiamin	oxythiamin	rats	FROHMAN and DAY [1949]
	riboflavin	araboflavin	rats	EULER and KARRER [1946]
	vitamin K	bis-hydroxy-coumarin	rats	LINK *et al.* [1943]
Pyrimidines	uracil	5-fluorouracil	mammal tumor	HEIDELBERGER *et al.* [1957]
Purine	guanine	8-azaguanine	tetrahymena	PARKS [1955]
	hypoxanthine	mercaptopurine	mammal tumors	HITCHINGS and ELION [1969]
Hormones	progesterone	pregnandiol	rats	HINSHAW and VELARDO [1951]
	thyroxine	3,5-diiodophenyl-carboxylate	rats	BARKER [1963]
Metabolites	acetate	fluoroacetate	rats	LIEBECQ and PETERS [1949]

1 See HOCHSTER and QUASTEL [1963] for a comprehensive treatise.

Michaelis-Menten and Lineweaver-Burk treatments provide the inhibition index, that ratio of the concentrations of I and M which produce a 50% inhibition of the initial reaction rate.

While competitive inhibition may be regarded as a contest between two slightly different molecules for a single locus on the active site of a macromulecule, noncompetitive inhibition usually allows the addition of both inhibitor and metabolite to the macromolecule. The addition of the inhibitor gives a different complex (RMI) which is less effective than the RM complex alone; the RMI complex cannot perform as fast as the RM macromolecule. Most noncompetitive inhibitors are allosteric; however, some bind at the active site of the macromolecule. Selected examples of antimetabolites are provided in table I. Specific information was taken from BARKER [1963], EULER and KARRER [1946], FROHMAN and DAY [1949], HEIDELBERGER *et al.* [1967], HITCHINGS and ELION [1969], LIEBECQ and PETERS [1949], LINK *et al.* [1943], McILWAIN [1941], PARKS [1955], PORTER and MEYERS [1945], SHUGAR [1969], WEISBURGER and SUHRLAND [1955], and BRIGGS *et al.* [1943]. The older concepts have been well explored [WOOLLEY, 1952] and categorized [HOCHESTER and QUASTEL, 1963]. Modern views are summarized in texts of biochemistry, pharmacology, medicine or toxicology. ALTMAN and DITTMER [1973] provided a good listing of specific antimetabolites to vitamins, amino acids, purines, pyrimidines and a few metabolites; this volume also collates other natural toxicants.

3. Food Toxicants

Food toxicants represent an environmental hazard which is much greater for some species than for others; the earthworm must process veritable mountains of soil for its sustenance while internal parasites feed efficiently from their host with little opportunity for ingesting harmful quantities of toxicants. Some natural foods contain a variety of natural substances in harmful quantities which differ little from the nonintentional toxicants, such as pesticides, which accumulate in food chains. The ubiquitous crescendo of pesticides and industrial wastes is testing the survival of many species on earth.

Potential food toxicants change with the evolution of earth and its peoples. Man's different cultures present problems which range from iron toxicity resulting from drinking fermented liquors made in iron pots to iron deficiency caused by eating excessive amounts of purified foods. Excessive cooking or other processing may destroy essential nutrients, thus increasing the content of biologically active but nonnutritive components. Foods are fortified to compensate for the loss of essential nutrients during processing and their preservation; food additives constitute a group of compounds which may be potentially toxic depending upon dose and circumstance. Food additives are normally present in nontoxic quantities and may be categorized as intentional or nonintentional food additives from an anthropocentric viewpoint [LUCKEY, 1972]. The 2,000 ± intentional food additives are problems of, by and for man. The use of these additives must be regulated to maximize the continuity of man's existence on earth.

The food toxicants least avoided are those inherent in natural foods. The nonintentional additives in the food of man and domestic animals can

Table II. Food toxicants

Class	Example
I. Inherent toxicants	
A. Mineral and antimineral	
1. Precipitable ions	Fe, Ca, Mg, Zn
	PO_4, OH^-, oxalate, high molecular weight, fatty acids
2. Co-precipitants	Ca → Zn salt ↓
	Zn → Cu salt ↓
3. Chelators	phytic acid, oxalic acid
4. Adsorbants	gums, undigested starch, cellulose
5. Others	radionuclides radiation, Pb, Se, As
	NO_2^-, NO_3^-, Fl
B. Protein, antiprotein	
1. Enzyme inhibitors	antitrypsin, widespread
a) Sorghum	antiamylase in bars
b) Potato poisoning	choline esterase (solanine)
c) Thistle	oxidative phosphorylation
2. Others	lectins (phytohemogluttinins)
	allergens, gluten
C. Antivitamin	
1. Enzymes	thiaminase
2. Other protein	avidin
3. Antagonists	dicumerol, antithiamin
a) Wheat, corn	niacin antagonist, niacytin
4. Others	metal ions adsorbed on the protein changing its conformation whereby the protein acts as an antigen
D. Biologic toxins	
1. Bacterial	photosensiting agents, polycyclic hydrocarbons, staphylococcus enterotoxins, ethionine, nitrosamines, *C. botulinum* toxins
2. Fungal	ergot, many mycotoxins (aflotoxins)
a) Amanita mushroom	toxic peptides
3. Plants	carcinogenic alkaloids and glucosides
a) Many plants	phytates, oxalates, cyanogenic glycosides and phenols (cathartics, gossypol and tannins), nitrosoamines, thiourea, opiates, xanthine, stimulants, tyramines, polycyclic aromatic hydrocarbons, carrageenan

Table II (continued)

Class	Example
b) Nutmeg xs	
c) Broad bean, its pollen	flavin hemolysin, pyrimidines
d) Legume beans	phaseolotoxin
e) Lathyrus	lathyrogens, aminonitriles
f) Grasses and grains	seleno-amino acids
E. Other toxicants	
1. Vitamin A	polar bear liver
2. Oxidants	
3. Plant toxins from animals	
a) Quail from hemlock seeds	hemlock
b) Mussels from dinoflagellates: saxitoxin	mytilotoxin
c) Shellfish poisoning	protista 'blooms'
fish poisonous from	ciquatera toxin
'food chain'	tetrodotoxin
'Puffer' fish	
d) Potatoes, green	solanine
4. Oncogens	
5. Allergens	
6. Oleander	alkaloids
7. Honey	andromedols and tutins
8. Goitrogens	brassica toxins and SCN
II. Nonintentional food additives	
A. Heat products	
1. Burnt meats	aromatic carcinogens
B. Products of harvest	
1. Pesticides	chlorinated hydrocarbons, herbicides
2. Fertilizers	dirt
3. Growth promotants	antibiotics
C. Storage materials	
1. Biologic products	aflotoxins
2. Growth suppressants	pesticides
D. Processing materials	polyethylene glycols, koleate
E. Packaging	
1. Plasticizers	diethyl phthalate, stearyl citrate
2. Paper products	'wax', rubber, Cr glycine complex

Table II (continued)

Class	Example
F. Accidental excesses	
1. Ink, Paint, glazing	lead
2. Pans	Fe, Cu, Zn, Sb, Sn
3. Wine	Fe, Cd, Pb
4. Beer	As, Fe, Co
III. Intentional food additives	
A. Nutritive	
1. Minerals	I^-, Fe^{++}, Fl^-
2. Vitamins	vit. A, vit. D
3. Amino acids	protein supplements
B. Storage	
1. Antimicrobic compounds	antibiotics, propionate, Na benzoate, $SO_2^=$, NO_2^-
2. Pesticides	
3. Aging compounds	
C. Processing	
1. Physical state modifiers	glycocolic acid, Na citrate, propylene glycol, gums
2. Flow properties	Mg silicate
3. Chemical state modifiers	
a) Moisture	Co tallate, isopropyl citrate, Na thiosulfate
b) pH	Ca citrate, lactic acid
c) Antioxidants	propyl gallate, nordihydroguaiaretic acid
D. Sensory	
1. Flavoring	
a) Aroma	plant extracts, musk
b) Taste	saccharine, spices
2. Color	

This table is predicated upon the handful of fine reviews listed in the accompanying text.

be partially controlled by proper food production management. Foods with excess radionuclides will be discarded; food containing known toxicants are avoided; heat processing will denature and inactivate most of the toxins; and normal cleanliness procedures will avoid or diminish the intake of many food toxicants. J. WEATHERSTON [in ALTMAN and DITTMER, 1973] lists over 300 insect produced toxicants; any of these could be found in unclean foods. Small quantities of pheromones and other semiochemicals are less important. An admirable listing of animal, plant and myco toxins is presented by ALTMAN and DITTMER [1973]. The limited biosynthetic potential of animals makes animal produce more constant in chemical composition than are plants. Fewer natural toxicants are provided to carnivores than are provided to herbivores.

Recent treatises on food toxicants for man and domestic animals allow a brief review of only the major features herein. The classification of toxicants (table II) places pesticides in the nonintentional category because the species which eat the food had no direct part in placing the additive in the food. BICKNELL [1961] presents a broad view of the dangerous chemicals in the food of man and animals. AYRES *et al.* [1962] emphasize the microbic aspect of food hazards and AYRES *et al.* [1968] presented a broad view of the safety of foods. OKE [1970] reviewed the toxicants in Nigerian foodstuffs. GONTZEA and SUTZESCU [1968] have reviewed natural antinutritive substances in foods for man and animals, and the Committee on Food Protection [1973] reviewed specific naturally occurring food toxicants. More details regarding toxicants in animals and in plants were presented by LIENER [1969, 1974]. Methodology and functional aspects of food toxicants were evaluated by ROE [1970]. Intentional food additives are presented by FURIA [1972]. This overview suggests that the number of hazards in foods is high and that the toxicants in food are potent. However, the large and varied biomass on earth suggests that the quantity of most food toxicants is too small to affect in any decisive way the majority of animal species which now contribute their full share to future evolution.

4. Other Toxicants

Consideration of essential nutrients, classic antimetabolites, food additives and inherent food toxicants suggests the existence of common denominators for other toxicants. Unfortunately, they exist only on most broad bases. A great number of ingested toxicants are taken in small amounts during each lifetime. The harmful effects might lead to chronic toxicity and may be most difficult to relate to any single toxicant. SCHROEDER [1972] revealed the recondite toxicity of several heavy metal salts only after feeding small quantities for more than one generation of rodents. Greater quantities of such compounds would evoke faster and usually more dramatic harmful responses. Some ingested toxicants resemble drugs taken for a specific malady; however, a review of therapeutic drugs is inappropriate here.

Since all ingested compounds are harmful in excess, it is important to differentiate the harmful action of low doses of compounds for different animal groups. This suggests two major categories: (a) primary toxicants affect the essential life processes of all animals, and (b) secondary toxicants affect phylogenetically acquired structures or functions.

The cell theory of SCHEIDEN and SCHWANN was a presage of more than the basic anatomical structure for all biology. The basic concepts for physiology, nutrition, biochemistry and toxicology are encompassed in the complex diversities exhibited by single celled organisms. As PROSSER and BROWN [1961] note, 'at the cellular level all organisms have more in common than in difference...'.

Those toxicants which affect the essential life processes of most animals include compounds which affect respiration (HCN), reproduction (actinomycin D), movement (curare), energy utilization (anti-niacin), membrane permeability (Ca chelators), and other specific characteristics of all life. Comparative toxicology of this group is based upon differential absorption, metabolism, hemostasis, and detoxication mechanisms of the animals exposed to these compounds. The complexities of intake for all materials in comparative toxicology are diagrammed in figure 1. The storage of the toxicant and/or its metabolites provides a basis for delayed action whenever the storage depot (fat) is being utilized. Another important variable is previous development of induced enzymes which metabolize the toxicant, i.e. the mixed function oxidases may effect changes in many different molecules other than the one which induced them. The opposite effect is observed from toxicants that cause liver damage and thus prevent or depress enzyme synthesis. These are examples of interactions which occur between toxicants and target tissues. The ability of HCN (from cyanophoretic glycosides) to penetrate most cells and the low capacity of animals to neutralize HCN make this a compound of limited interest in comparative toxicology. It presents a valid test for the comparative biochemist to detect those invertebrate species which can survive without a functioning cytochrome system.

The 'essential process' toxicants differ from differential toxicants when phylogenetically developed physiologic and biochemical activities make different species of animals react differently towards the same toxicant. Toxicants which affect phylogenetically acquired structures or functions present a systematic approach to comparative toxicology.

II. Phylogenetic View

The unity of comparative nutrition is assured from the phylogenetic continuity of histology, anatomy, physiologic functions and biochemic pathways. The same basis provides a unity in comparative toxicology. Materials which are toxic to an essential process in one cell are predictably toxic to other cells having the same metabolic pathway. New biochemic pathways often utilize components of previously established metabolic sequences. The repetitious nature of major features of comparative metabolism is evident in the presentation of a single biochemic reaction for different phyla, the treatise on comparative nitrogen metabolism [CAMPBELL, 1970] illustrates this unity.

Diversity allows one animal's meat to be another animal's poison. The diversity in functional capacity and in anatomic structure for a single function underscores significant contrasts for comparative toxicology. Contrasting with the impressive unity of comparative animal nutrition, the diversity of metabolic functions provides much fuel for comparative toxicology.

It remains a challenge to relate comparative toxicity to differences in anatomic structure, physiologic function, or metabolic capability. Phylogeny evokes subtle but real differences in organisms based upon newly developed parameters which are not found in previous life forms. A function is embellished or is fulfilled by a new structure, a new compound or a new reaction sequence. Even here, phylogenetic changes bow to the unity of comparative biochemistry; a new reaction sequence or the production of a new compound requires new enzymes which are susceptible to the many poisons which affect other enzymes. The specificity of enzymes is due to secondary, tertiary or quaternary structure more than to amino acid composition. Excepting amino acid antagonists, important biocides for selective toxicology may be expected to relate less to enzymatic composition than to phylogenetically new metabolic compounds (i.e. hormones), reactions and control mechanisms for physiologic functions of the new systems. Primary toxicants are those which are associated with biochemical unity of all animals, secondary toxicants are those which act on systems or structures developed during phylogeny. Secondary toxicants should be harmful only to

Table III. Comparative membrane permeability in living cells

Compound	Permeability (moles × 10^2/sec/μm^2/molar conc. difference)		
	protozoa (Gregarina)	marine animal (Arbacia egg)	ox erythrocyte
1,2-Dihydroxypropane	13,200	13,000	4,000
Glycol	6,700	7,300	2,100
Urea	2,500	–	78,000
Glycerol	180	50	17

Taken from ALBERT [1973].

Table IV. Inhibition (%) of phosphofructokinase

M concentration	Antimony potassium tartrate		Stibophen	
	trematode	rat brain	trematode	rat brain
1×10^{-3}	100	32	100	0
3×10^{-4}	100	0	85	0
3×10^{-5}	32	0	0	0

Taken from MANSOUR and BUEDING [1954].

those phyla which carry the new system or structure and should be less harmful to organisms in lower phyla.

A simplified view of the phylogenetic basis for comparative toxicology is presented in table III of the introductory chapter. Important features which may be, but rarely have been, utilized for selective toxicity relate to new essential metabolites, functions, or structures which are present in only part of the animal groupings. Salient examples include: type of phosphogen, iron hemoglobin, complex nerve synapses, central nervous system, complex digestive system, cardiovascular system, mammary system, endocrine products and target cells, blood-forming tissues, erythrocytes, excretory functions or organ, defense systems, complex cell differentiation and temperature control.

Each morphologic or physiologic system exhibits variations which reflect both phylogenetic and ontogenetic development. ALBERT [1973]

Table V. Comparative toxicity of insecticides (acute LD_{50}; ALTMAN and DITTMER [1968])

Action Name	Mammalia	Aves[1] (oral), ppm	Insecta (topical, adult)	Insecta (larva, media)	Teleosts LD_{100}[2], ppm	Teleosts LD_{50}, ppb	Crustacea LD_{50}, ppb	Insects LD_{50}, ppb
'Contact'								
Chlordane	180–750	331–858			0.02–0.08	22–57	4–40	15
Neurotoxin								
Allethrin	680		8	0.14				
Phthalthrin	>20,000		2	0.06				
Anticholinesterase								
Azinphosmethyl	11	75–2,000	2.7	0.03		5–93	0.1–21	1.5–12
Carbaryl	850	>5,000	900	1		9		
Dizinon	76	47–245	3	0.09		6	1–200	1.7–25
Fenitrothion	500	157–2,482	2	0.006				
Malathion	1000	2,128–>5,000	28	0.08	12	120–13,000	0.8–180	1–10
Parathion	3.6	44–365	0.9	0.003	1–3	65–1,410	0.4–3.5	1.5–4.2
Propoxure	104		22	0.33				
Peripheral neurotoxin								
DDT	150–600	311–1,869	2	0.007	0.1–0.5	2–19	0.4–4	2–7
Methoxychlor	6,000	>5,000	9	0.07	0.05–0.1	7.5–66	0.8–5	0.6–1.4

Central neurotoxin and hepatotoxin								
Deldrin	40–90	39–185	1	0.008	0.006–0.25	6–16	5–140	0.5–39
Heptachlor	60–225	92–480	2	0.006	0.06–0.25	17–59	2–47	0.9–2.8
Lindane	40–200	425–>5,000	0.9	0.025		2–87	10–520	4.5
Aldren		92–894				8–46	8–9,800	1–200
Endrin		14–22				0.5–1.2	0.4–20	0.3–2.4
Toxaphen		538–838				2–18	6–28	1–3
Herbicides								
Dalapon		>5,000				3,000,000	11,000–16,000	
Dichlobenil		1,500–>5,000				20,000	3,700–34,000	7,000–20,700
Diquat		1,346–>5,000				7,800–35,000	48	100–33,000
2-4,D(BEE)		>5,000				5,600	440–5,900	1,600
Fenac						15,000	4,500–12,000	55,000
Allethrin						19–56	8–56	2.1

1 Heath *et al.* [1972].
2 Fish data from Johnson [1968] and US EPA [1973]. Effects of pesticides in water.

provides a comparison of the permeability of different membranes to a variety of compounds (table III). The differences are greatest for physiologically important compounds. ALBERT presents superb discussions about the sparse information on the use of different metabolic pathways by different species. Selective toxicity of certain toxicants have practical significance; MANSOUR and BUEDING [1954] illustrate this selective toxicity as the basis for the use of antimonials in the treatment of schistosomiasis in man (table IV).

Although the metabolism of a toxic material is usually considered in terms of detoxication, some compounds become more harmful during metabolism. KRUEGER and O'BRIEN [1959] provide an example of this as a factor in selective toxicity; insects readily convert malathion to the more toxic derivative, malaoxon, while mice do not.

Biochemical unity provides a practical basis for the use of a wide variety of animals, including protozoans, in the search for general poisons. All phyla have the capacity to absorb soluble compounds, and possess digestive, glycolysis and TCA enzymes, electron transport system (some parasites have lost the terminal enzymes), oxidases, sensory systems and at least a degree of sexuality and mitosis with DNA, RNA and protein synthesis involvement. Species-specific poisons have not been well developed and will be difficult to develop due to the sporadic occurrence of compounds or systems apart from phylogenetic developments. For example, hemoglobin and acetylcholine have been reported in protozoans; therefore, toxicants related to these compounds could be harmful to animals which do not have blood or a nervous system, respectively.

The following phylogenetic review of toxicants emphasizes the different reactions obtained throughout the continuum of dose-response relationships. Wherever possible, phyla or class-specific toxicants will be used as examples. Differences between phyla will be examined in order to reveal toxicants for a specific phylum and all higher organisms having that specific character and capability; since toxicity is based upon a characteristic not found in lower forms, such compounds should be relatively nontoxic to lower forms. Unfortunately, the visitudes of economic pressure and lack of facilities for thorough comparative study retard development of this grand vista of potential knowledge. Available information in this area has been summarized in table V. This summary indicates how little actual information can be suggested under the column entitled 'action'. Although the insecticide is presented to the different groups of animals in different ways, the major effect on the most susceptible target tissue remained the same. WINTERINGHAM [1969] reviewed selective toxicity of insecticides. Organic phosphates inactivate enzymes by phosphorylating them, e.g. acetocholine esterase. The organic carbamylates inactivate by carbamylation in like manner. Other actions of pesticides, such as a delayed neurotoxicity by organic phosphates were reviewed by KAY [1973]. PIMENTAL [1971] provides a good collation of comparative information on pesticides, and the interactions between pesticides and wildlife were presented by COPE [1971]. LITTLE [1970] collated toxicity data of water pollutants for mammals and the NAS Committee on Water Quality Criteria [1974] collated data on minerals and selected toxicants that act on fish and some crustacea.

A. Protozoa

Although the protozoa show no real cell differentiation within species and little potential for complexities of intercell communication on a permanent basis, many protozoas are large and highly developed. A single cell may be very complex and show a variety of movements and specialized functions. Some protozoa have mitochondria as well as most organelles exhibited by eucaryotic cells. They may have cilia for obtaining food and may produce an acid pH for enzymic hydrolysis in food vacuoles. The general problems of osmotic pressure, membrane permeability, food utilization, excretion, and metabolism make protozoa good representatives of all animals. Thus, systems of toxicty and mechanisms of action have been worked out for protozoa which are applicable to higher forms of life [HUTNER and KITTER, 1955; DEROW, 1965]. Information on nutritional requirements, antimetabolite action and food toxicants for protozoans is applicable to most metazoans.

The disasterous effects of many parasitic diseases in man and domestic animals has made Protozoa one of the most studied of all phyla. The variety of toxic materials listed in table VI for protozoa is not necessarily specific for protozoa, but their large cell size, high metabolic rate and the relatively high permeability of their cell membrane makes these animals more susceptible than most host animals for this group of drugs. The antibacterial drugs are highly effective in protozoan diseases for the specific reason that the protozoa are adapted to the intestine of animals which harbor myriad bacteria. As PHILLIPS [1968] has shown, the maintenance of anerobic conditions in the intestinal tract is a prerequisite to amoebic pathogenicity. Bacteria, such as *Bacillus subtilus,* are required to potentiate the virulence of

Table VI. Examples of selective biocides

Phyla grouping	Material or compound	Action
Protozoa	antibiotics	bactericidal to some essential symbionts; effective in amoebiasis
	antimalarials: quinine	blocks respiration, both physiologic and biochemical
	pyramethamine	gametocidal
	chlorguanide	gametocidal
	quinacrine	flavin antimetabolite
	chloroquine	
	amodiaquin	
	arsenicals: pentavalent	reduced to trivalent
	trivalent, arsenoso	trypanocidal
	trivalent, arseno	oxidized to active compounds
	antimonials, trivalent	inhibit sulfhydryl enzymes
	stilbamidine	denatures nucleic acid, inhibit O_2 consumption
	metronidazole	block ferredoxin
Platyhelminthes	mepracrine	
	dichlorophen	inhibits phosphorylation, ATP inhibited
	chlorsalicylamide	inhibits phosphorylation, ATP inhibited
	niclosamide	inhibits phosphorylation, ATP inhibited
	antimonials	inhibits phosphorylation, ATP inhibited
	carbon tetrachloride	inhibits phosphorylation, ATP inhibited
	stibophen	inhibits phosphorylation, ATP inhibited
	lucanthene	inhibits phosphorylation, ATP inhibited
Nematoda	piperazine	inhibits succinate anabolism and blocks acetylcholine action
	tetramisole	inhibits succinate anabolism and blocks acetylcholine action
	suramin	inhibits succinate anabolism and blocks acetylcholine action
	diethylcarbamazine	whipworms
	viprynium	thread worms
	thiabendazole	larvacidal at 10^{11} dilution
	tetrachloroethylene	hookworm
	bephenium-HO-naphthanoate	myoneural block
	hexaresorcinol	muscle paralysis
	stilbazium iodide	block glucose transport, reduced egg capacity

Table VI (continued)

Phyla grouping	Material or compound	Action
	trichlorophenal	block glucose transport, reduced egg capacity
	piperazine	blocks acetylcholine at neurotransmission
	proteolytic enzymes	digest cell structures
Arthropods	chlorinated hydrocarbons	
	DDT	inhibits lactic dehydrogenase
	mirex	inhibits lactic dehydrogenase
	MYL	ovicide
	rotenone	inhibits NADH dehydrogenation
	pyrethrum	
	organophosphates	blocks acetylcholine esterase
Snails	acrolein	also kills fish
Mammals		
Rodenticides	arsenic trioxide	degeneration of liver and kidney tissues
	coumarin	hemorrhage
	strychnine	anoxia from muscle stricture
	Zn_3P_2	convulsions from PH_3
	Tl_2SO_4	neuritis and convulsions
	$BaCO_3$	muscle paralysis
	Na fluoracetate	convulsions
	pival	
	ANTU	
	chlorinated HC	interrupts estrus

Entamoeba histolytica. Nitrogen heterocyclic trichomonicides seem to be selective toxins for these flagellates due to the fact that these compounds (i.e. metronidazole) may block hydrogen transport by ferredoxin, a coenzyme with little significance in mammals [ALBERT, 1973]. For other drugs in this series the selective toxicity results primarily from the poor absorption by the host. Therefore, the drug can be given orally and affect the microbes in the intestines without materially damaging host tissues; this concept will be applicable for other drugs and other intestinal parasites.

Little is known of the few species which constitute the phylum Metazoa.

B. Diploblasts

Toxicology of diploblast phyla does not appear in texts. Certain coelenterata have been used extensively in some areas of nutrition; for example, the establishment of glutathione as an appetite factor for hydra. The control of sponge predators is important for certain Pacific islands.

Porifera provide cell differentiation, specialization, intercell communication and coordination of different kinds of cells within one individual. In general, the endoderm serves alimentation purposes while the ectoderm serves defense, sensory and food gathering purposes. Digestion is intracellular with ameboid cells participating. These functions are altogether in protozoans and provide acceptable bases for differential toxicants.

Coelenterata are more complex than are Porifera. They show both budding and bisexual reproduction and have a single opening for the alimentary tract. Their specialization provides contractile cells, sting cells and sensory cells as well as a nerve network. Digestion is still mostly relegated to ameboid cells. Perhaps different toxicants could be based upon sex semiochemicals which would not affect more primitive animals; the specificity of pheromones is used to betray noxious animals into a diversity of pitfalls.

Ctenophora animals are slightly more complex than are coelenterata. Their nerve system is more organized. They have organized locomotor organs and are hermaphroditic. No effects of specific toxicants were found for these animals.

C. Acoelomates

This group of worms is well known through representation of parasitic forms in man and domestic animals. They may have no alimentary openings and simply receive materials by diffusion through their cell membranes. Food is ingested and digested intracellularly. Cephalic orientation is well defined with a primitive brain, and some have primitive fluid vessels and eyes. They seem to have limited osomo-regulation. However, some of them do have hemoglobin according to MANWELL [1960]. Primitive sensory organs are found and histidine appears to be functioning in the cell economy. Many of the drugs used against platyhelminths (table VI) depress energy utilization. This series of drugs generally is found to inhibit phosphorylation and utilization of phosphogens.

D. Pseudocoelomates

The best known representatives of this group of phyla are the nematodes. These triploblasts show good utilization of the mesoderm. They have a circulatory system to increase dispersion of oxygen and nutrients; this allows greater potential for osmoregulation and a greater body thickness than is seen in acoelomates. The alimentary tract has a one way flow with a well-defined mouth and anus. They are often divided into two sexes. They have a simple excretory system and some capability of osmotic balance with chloride ion being active in excretion processes. Hemoglobin has been found in the body fluid and body wall. As indicated on table III of the introduction, histamine is active, hermaphroditism is continued, urea and ammonia excretion is usual. Many nematodes are of medical interest and representative drugs used against them are presented in table VI. These drugs tend to be primary toxicants which would inhibit the cellular production of energy; presumably the host escapes harmful action of these drugs due to the differential in the amount absorbed for any given dose in the intestine. Selective toxicity against round worms is exhibited by tetramizole [ALBERT, 1973] which has little activity in protozoa, flat worms or bacteria.

When the complete biologic activity spectrum of antibiotics was examined, chloramphenicol was the most consistent or promotant for *Turbatrix aceti* [LUCKEY, 1963a]. The effect of two antibiotics upon colony size (growth, reproduction and survival) is shown in figure 8. When subjected to severe heat stress, the vinegar eels showed lower mortality in the presence

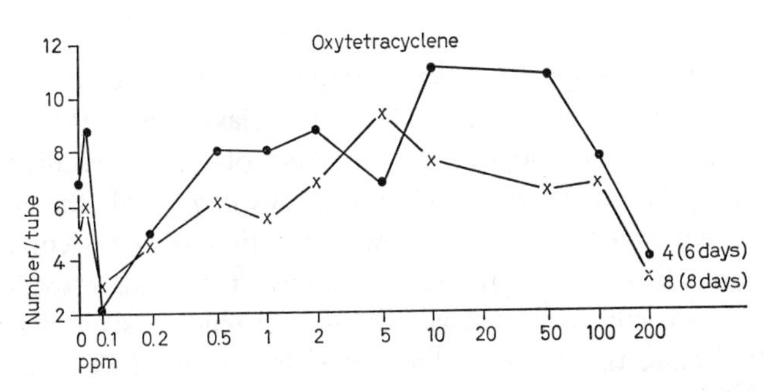

Fig. 8. Effect of low levels of antibiotics upon reproduction and/or survival of the vinegar eel (*Turbatrix aceti*). Note the consistent stimulation with the lowest dilution. From LUCKEY [1963a].

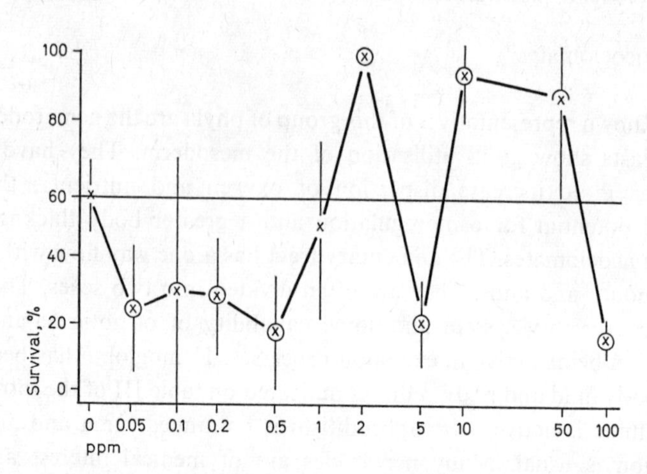

Fig. 9. Effect of different levels of staphylomycin upon the survival of the vinegar eel *(Turbatrix aceti)* to heat stress. The vertical line = 2 SD. The probability for the comparison of experimental with control is indicated by the completeness of the circle about the point representing the mean of the experimental group: ◯, p = < 0.001; ◯, p = 0.001–0.01; ⌒, p = 0.011–0.02; ◝, p = 0.021–0.05. From LUCKEY [1963b].

of low doses of staphylomycin, chloramphenicol, procain, penicillin, streptomycin or oxytetracyclene [LUCKEY, 1963b]. The complex types of dose-response curves obtained with staphylomycin and chloramphenicol are most intriguing (fig. 9). As anticipated, the high doses were toxic.

E. Schizocoela

Insecticides have been applied extensively to control the propagation of members of this group of phyla. Excepting a few classes of Arthropoda, very little data has been catalogued for members of this super division [NEGHERBON, 1959]. Often the action of insecticides is general; therefore, for any given dose level the relatively low absorption of these complex organic compounds by other arthropods and domestic animals provides a basis for selective toxicity. Animals with fat depots tend to sequester the lipid soluble toxicants, this provides limited protection until the fats are heavily utilized for energy. Available data on comparative toxicity of insecticides is summarized in table V. The insect larva is highly sensitive to many of these insecticides, irrespective of their mode of action, while the

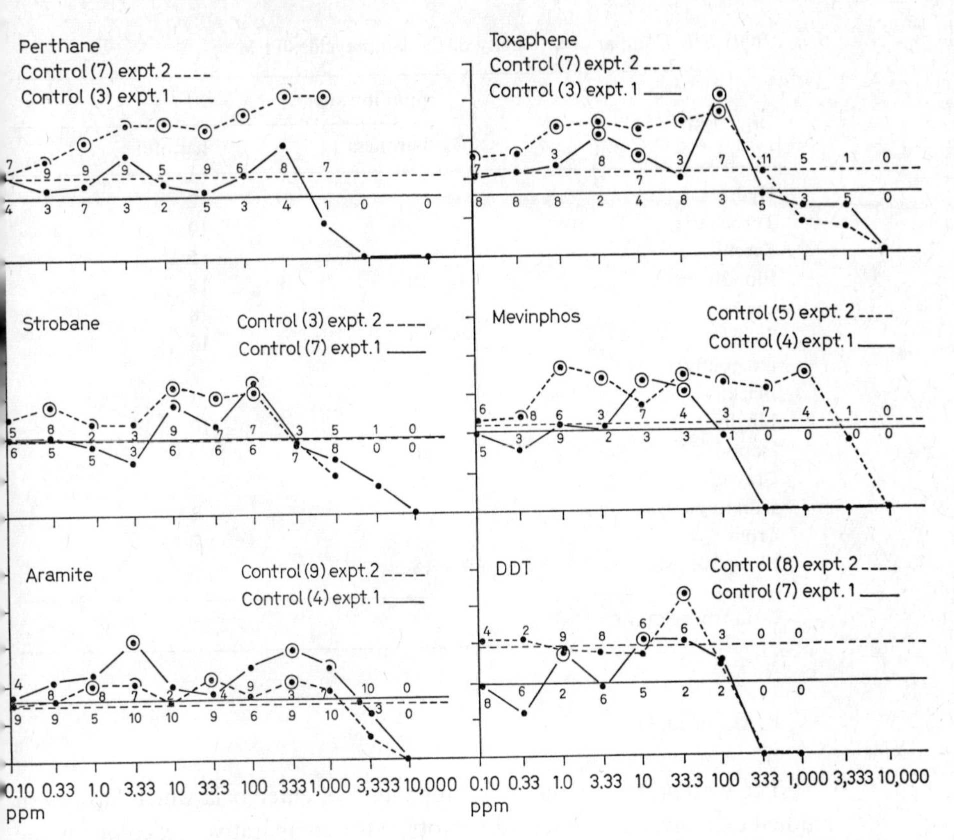

Fig. 10. Activity spectrum of insecticides in crickets *(Acheta domesticus)*. The statistical significance is given by the completeness of the circle about each point as defined in figure 9. From LUCKEY [1968].

adult insect is relatively immune. There appears to be another safety factor between the sensitivity of adult insects and the vertebrates. When the complete biologic activity spectrum of insecticides was examined (fig. 10), low doses of most ingested insecticides stimulated the growth of insects [LUCKEY, 1968]. These data confirmed the prediction of the major thesis of hormology [LUCKEY, 1959).

Selective molluscicides include N-tritylmorpholine, amides, such as dichloronitrosalicyl amilide, and a glycoside secreted by *Phytolacca dodecandra* (an Ethiopian water plant); these are relatively harmless to other aquatic animals or mammals [ALBERT, 1973].

Table VII. Comparative toxicity of the lampreycide, TFM

Organism	ppm in water	
	harmless	harmful
Hydra		2
Turbellaria		10
Leech		5–8
Bloodworm	10	15
Clam		8
Snail	10	15
Dragonfly	20	
Blackfly		3
Mayfly		10
Isopod	20	
Crayfish	20	
Lamprey larvae		2
Trout		7–13
European fish	18	

Data from SCHNICK [1972].

F. Entercoela

The vertebrates are the only subphylum of entercoela which have been studied extensively. A practical prototype for comparative toxicology is the development of selective toxicants to control the sea lamprey which invaded the US Great Lakes. Of the handful of compounds which were acceptable [SCHNICK, 1972], 3-trifluoromethyl-4-nitrophenol (TFM), a halogenated nitrophenol was chosen because it killed all lamprey larva within a short time. Comparative toxicity studies (table VII) indicate that coelenterates are as sensitive as lampreys and the ratio of toxicity of TFM for lampreys and a variety of aquatic forms, including fish, is very narrow. It seems possible that factors of mixing, temperature, water composition and food supply may make the toxicity nonselective. It is too soon to evaluate nature's response to this type of toxicologic control of aquatic ecology.

The cohabitation of certain rodents with man through millenia has made it necessary to develop effective rodenticides [FEST and SCHMIDT, 1973; HEATH *et al.*, 1972; JOHNSON, 1968; STREET, 1972]. The list of available compounds (table VI) suggests that most are primary toxicants which are selective to rodents only by appropriate placement. These rodenticides and other similar toxicants have become notorious by their occasional misuse which results in trauma or death to domestic animals and man. Genus or family specific toxicants will be very rare until biocides are based upon immune responses. ALBERT [1973] notes a few species or class specific reactions. Only man and old world primates dehydrogenate quinic acid to benzoic acid; aromatic amines are acetylated by most vertebrates with frogs, turtles and dogs being the exception; and rats are the only mammals susceptible to norbormide. He also notes that man can tolerate doses of HCN and strychnine which are lethal to dogs and rabbits, respectively. However, man is 15 times more susceptible to atropine than are rabbits.

III. Adaptation

The changing environment provides direction and impetus for evolutionary development. A short generation time is compatible with rapid adaptation via mutation and/or selection to provide populations resistant to specific ingested toxicants; this has been well studied in insect adaptation to insecticides [WINTERINGHAM, 1965]. Since the rate of mutation for two separate adaptations is appriximately the product of their individual rates, the simultaneous appearance of two distinct poisons provides a barrier which may effectively block adaptation in susceptible species. Practical experience with insect control using DDT shows that species adaptation to this drug provide resistance to similar poisons (TDE, and methoxychlor) but not to members of a second group of insecticides; yet resistance to one member of the second group assures resistance to all members (aldren, dieldrin, lindane, chlordane, heptachlor and toxaphen). SMITH [1964] reviewed insect detoxication mechanisms. Similar patterns of resistance are seen in mammals via inducible enzymes and are discussed in modern pharmacology texts. ALBERT [1973] notes that the insecticide rotenone is highly toxic to isolated mitochondria of both fish and mammals; fish susceptibility and mammalian resistance to rotenone is based upon the differences in metabolic oxidations of such toxicants.

If survival of the species depends upon two separate mutations for new enzymes to metabolize two different toxicants, the chance for proper adaptation reaches the limits for reproductive capacity of any species. The number of individuals which possess the survival mutations within the habitat must be in sufficient density to continue competitive reproduction. RUSSEL [1960] recommended addition of antimalarial drugs to salt to destroy gametocytes in the human host and simultaneous dispersal of insecticide to reduce the mosquito population for more effective control of endemic malaria. This one-two punch of the host-parasite would appear to be less effective than the double punch of two simultaneous poisons directed to the species to be eliminated.

The appearance of new ingested toxicants in the nurture of the animal is balanced by a series of protective mechanisms: homeostatic elimination processes, detoxication via previously established metabolic sequences which include enzyme induction, decreased absorption, sequestering in tissue (lipid), changed eating habits (avoidance) through social action, human control of chemical pollution and, finally, mutation. Recent experiences suggest that most large animals have long generation times which are too great to allow the evolutionary development of resistant strains for the sudden influx of many new chemicals into the environment. As emphasized earlier, the quantity of new compounds presented in combination becomes a most important factor in determining toxicity for species survival; large animals need much time to develop protective systems via selective adaptation and/or mutation.

Summary

The complete activity spectrum of ingested toxicants indicates that many are stimulants when administered in very small doses. An excess of any nutrient is harmful and certain essential nutrients may stimulate growth both at subadequate or excess dose levels. This phenomenon is comparable to the growth stimulation obtained by feeding antibiotics or other growth promotants and is related to the stimulation of many functions by minute quantities of toxicants. This is hormology, the study of excitation. This blending of nutrition, hormology, and toxicology presents a unified concept of the intricacies by which ingested chemicals may cause opposite and diverse reactions depending upon the dose, the conditions under which they are used, and the character of the individual exposed to them.

Ingested toxicants include excesses of essential nutrients, specific antimetabolites to these nutrients, toxicants inherent in food, nonintentional food additives, intentional food additives, and other toxicants. Some toxicants affect an essential life process and thus would be harmful for most animal species. Other toxicants affect selective groups of animals depending upon their phylogenetically acquired characteristics. The data base for these two groupings is not strong.

Comparative toxicity is of major importance in the development of drugs and pesticides. Many of the most active compounds cannot be used because they are too toxic for the host or for a protected species. Selective toxicity may be due to poor absorption by one species, to more efficient sequestering or cumulation, metabolic action which may be activation or detoxification, excretion and gene mutation. The short generation times for many invertebrates allows resistant strains to develop.

The practical development of selective biocides by man has centered mostly on protozoa, platyhelminthes, nematodes, insects and rodents. Generally, these and the development of lamprey control chemicals are due less to the selective toxicity than to relatively minor differences in absorption, utilization or ability to activate or detoxicate compounds.

The phylogenetic character of the diversity of animal life should provide a good working basis for comparative toxicology. This presents a goal which has been less useful than was anticipated due to the unity of life. Organs with new morphologic, physiologic or biochemic characteristics are composed of the same basic building blocks which compose most other tissues. Selective toxicity must utilize drugs which act upon new physiologic functions, new metabolic pathways and the new metabolites which control those functions.

Consideration must be given to the ability of different animals to resist the appearance of new toxicants in the environment. Here the advantage for selective absorption, low metabolic rate, homeostatic excretion mechanisms, storage in fat depots, and enzyme induction resides with the large animals. However, these advantages do not equal the advantages enjoyed by small invertebrates of a short generation time and reproductive capacity for large numbers. The potential for mutation into resistant strains promises to outweigh any and all other advantages. Man must utilize his brain and social institutions to provide adequate understanding of the need to control the concentration of ingested toxicants in the total environment in order to best assure the continuity of a balanced ecology for man on this earth.

References

ALBERT, A.: Selective toxicity; 5th ed. (Chapman & Hall, London 1973).

ALTMAN, P. L. and DITTMER, D. S.: Metabolism (Federation of Am. Soc. for Experimental Biology, Washington 1968).

ALTMAN, P. L. and DITTMER, D. S.: Biology data book; 2nd ed., vol. 2 (Federation of Am. Soc. for Experimental Biology, Washington 1972, 1973).

AYRES, J. C.; BLOOD, F. R.; CHICHESTER, C. O.; GRAHAM, H. D.; MCCUTCHEON, R. S.; POWERS, J. J.; SCHWEIGERT, B. S.; STEVENS, A. D., and ZWEIG, G.: The safety of foods (Avi Publishing, Westport 1968).

BARBER, R. S.; BRAUDE, R.; MITCHELL, K. G.; ROAK, J. A. F., and ROWELL, J. G.: Further studies on antibiotics and copper supplements for fattening pigs. Br. J. Nutr. *11:* 70–79 (1957).

BARKER, S. B.: Thyroxine analogues; in HOCHSTER and QUASTEL Metabolic inhibitors, a comprehensive treatise, pp. 535–566 (Academic Press, New York 1963).

BICHNELL, F.: Chemicals in your food (Emmerson Books, New York 1960).

BRIGGS, G. M., jr.; LUCKEY, T. D.; TEPLEY, L. J.; ELVEHJEM, C. A., and HART, E. B.: Studies on nicotinic acid deficiency in the chick. J. biol. Chem. *168:* 517–522 (1943).

CAMPBELL, J. W. (ed): Comparative biochemistry of nitrogen metabolism, 2 volumes (Academic Press, New York 1970).

COHEN, P. A.; SCHNEIDMAN, K.; FELLNER, F. G.; STURMAN, J. A.; KNITTLE, J., and GUALL, G. E.: High pyridoxine diet in the rat: possible implications for megavitamin therapy. J. Nutr. *103:* 143–151 (1973).

Committee on Food Protection, MRS: Toxicants occurring naturally in foods; 2nd ed. (Nat. Academy of Sciences, Washington 1973).

Committee on Water Quality Criteria, National Academy of Science: Freshwater aquatic life and wildlife, EPA; 2nd ed. (US Government Printing Office, Washington, in press, 1974).

COPE, O. B.: Interactions between pesticides and wildlife. Annu. Rev. Entomol. *16:* 325–264 (1971).

DEROW, M. A.: The treatment of parasitic diseases; in BELDING Textbook of parasitology; 3rd ed., pp. 1263–1298 (Appleton Century Crofts, New York 1965).

Environmental Protection Agency: Effects of pesticides in water, 85 pp. (US Environmental Protection Agency, Washington 1973).

EULER, H. V. and KARRER, P.: Iso-alloxazinderivate als Antagonisten des Riboflavins. Helv. chim. Acta *29:* 353–354 (1946).

FEST, C. and SCHMIDT, K. J.: The chemistry of organophosphorus pesticides: reactivity, synthesis, mode of action, toxicity (Springer, Berlin 1973).

FROHMAN, C. E. and DAY, H. G.: Effect of oxythiamin on blood pyruvate-lactate relationships and the excretion of thiamine in rats. J. biol. Chem. *180:* 93–98 (1949).

FURIA, T. E.: Handbook of food additives; 2nd ed. (CRC Press, Cleveland 1972).

GONTZEA, I. and SUTZESCU, O.: Natural antinutritive substances in foodstuffs and forages (Karger, Basel 1968).

HEATH, R. G.; SPANN, J. W.; HILL, E. F., and KREITZER, J. F.: Comparative dietary toxicities of pesticides to birds. Special wildlife rep. Wildlife, No. 152, 57 pp. (1972).

HEIDELBERGER, C.; CHAUDHURI, N. R.; DANNEBERG, P.; MOOREN, D.; GREISEBACK, L.; DASCHINSKY, R.; SCHNITZER, R. J.; PLEVEN, E., and SCHEINER, J.: Fluorinated pyrimidines, a new class of tumor-inhibitory compounds. Nature, Lond. *179:* 663–666 (1957).

HISHAW, F. L. and VELARDO, J. T.: Inhibition of progesterone in decidual development by steroid compounds. Endocrinology *49:* 732–741 (1951).

HITCHINGS, G. H. and ELION, G. B.: Thiopurines as inhibitors of the immune response; in SHUGAR Biochemical aspects of antimetabolites and of drug hydroxylation, pp. 1–10 (Academic Press, New York 1969).

HOCHESTER, R. M. and QUASTEL, J. H.: Metabolic inhibitors (Academic Press, New York 1963).

HUTNER, S. H. and LWOFF, A. (eds): Biochemistry and physiology of protozoa (Academic Press, New York 1955).

JOHNSON, D. W.: Pesticides and fishes – a review of selected literature. Trans. Am. Fish. Soc. *97:* 398–424 (1968).

KAY, K.: Toxicology of pesticides. Recent Adv. Envir. Res. *6:* 202–243 (1973).

KRUEGER, H. R. and O'BRIEN, R. D.: Relationship between metabolism and differential toxicity of malthion in insects and mice. J. Econ. Entomol. *52:* 1063–1067 (1959).

LIEBECQ, C. and PETERS, R. A.: The toxicity of fluoroacetate and the tricarboxylic acid cycle. Biochem. biophys. Acta *3:* 215–230 (1949).

LIENER, I. E. (ed): Toxic constituents of plant foodstuffs (Academic Press, New York 1969).

LIENER, I. E. (ed). Toxic constituents of animal foodstuffs (Academic Press, New York 1974).

LINK, K. P.; OVERMAN, R. S.; SULLIVAN, W. R.; HUEBNER, C. F., and SCKEEL, L. D.: Studies on the hemorrhagic sweet clover disease. XI. Hypoprothombinemia in the rat induced by salicylic acid. J. biol. Chem. *147:* 463–474 (1943).

LITTLE, A. D. C.: Water quality criteria data book. Organic chemical pollution of freshwater for EPA, vol. 1., 379 pp. Stock No. 5501–0144 (US Government Printing Office, Washington 1970).

LUCKEY, T. D.: Mode of action of antibiotics – evidence from germfree birds. 1st Int. Conf. Use of Antibiotics in Agriculture, pp. 135–145 (Nat. Academy of Sciences, Washington 1956).

LUCKEY, T. D.: Modes of action of antibiotics in growth stimulation. Recent progress in microbiology, pp. 340–349 (Almquist & Wiksell, Stockholm 1958).

LUCKEY, T. D.: Antibiotics in nutrition; in GOLDBERG Antibiotics, their chemistry and non-medical uses, pp. 173–321 (Van Nostrand, Princeton 1959).

LUCKEY, T. D.: A study in comparative nutrition. Comp. Biochem. Physiol. *2:* 100–124 (1960).

LUCKEY, T. D.: Stimulation of *Turbatrix aceti* by antibiotics. Proc. Soc. exp. biol. Med. *113:* 121–124 (1963a).

LUCKEY, T. D.: Antibiotic action in adaptation. Nature, Lond. *198:* 263–265 (1963b).

LUCKEY, T. D.: Insecticide hormologosis. J. econ. Entomol. *61:* 7–12 (1968).

LUCKEY, T. D.: Introduction to food additives; in FURIA Handbook of food additives; 2nd ed., pp. 1–26 (CRC Press, Cleveland 1972).

LUCKEY, T. D.: Hormology as the scientific basis of homeopathy. 3rd Int. Symp. Biological Medicine, Lausanne (in press, 1974).

LUCKEY, T. D.: Hormology with inorganic compounds. Environ. Qual. Safety *1:* suppl., pp. 81–103 (1975).

LUCKEY, T. D. and STONE, P. C.: Hormology in nutrition. Science *132:* 1891–1893 (1960).

MCDANIEL, E. G. and DAFT, F. S.: Effect of ascorbic acid in prevention or delaying multiple B_1 vitamin deficiencies. Fed. Proc. Fed. Am. Socs exp. Biol. *13:* 468 (1954).

MCILWAIN, H.: Bacterial inhibition by aminosulphonic analogues of some natural amino carboxylic acids. Br. J. exp. Path. *22:* 148–155 (1941).

MANWELL, C.: Comparative physiology. Blood pigments. Annu. Rev. Physiol. *22:* 191–244 (1960).

MANSOUR, T. E. and BUEDING, E.: The actions of antimonials on glycolytic enzymes of *Schistosom mansoni.* Br. J. Pharmac. Chemother. *9:* 459 (1954).

NEGHERBON, W. O.: Handbook of toxicology. Insecticides, vol. 3 (Saunders, Philadelphia 1959).

NEVILLE, P. F. and LUCKEY, T. D.: Carbohydrate and roughage requirement of the cricket. *Acheta domesticus.* J. Nutr. *78:* 139–146 (1962).

OKE, O. L.: Toxic chemicals in Nigerian foodstuffs. Indian J. Nutr. Dietet. *7:* 119–123 (1970).

PARKS, R. E., jr.: Antimetabolite studies in tetrahymena and tumors; in RHOADS Antimetabolites and cancer, pp. 175–197 (Am. Association for the Advancement of Science, Washington 1955).

PHILLIPS, B. P.: Induced amebic enteritis in the germfree, monoinoculated, and conventional guinea pig; in MIYAKAWA and LUCKEY Advances in germfree research and gnotobiology, pp. 279–286 (CRC Press, Cleveland 1968).

PIMENTEL, D.: Ecological effects of pesticides on non-target species, 220 pp., No. 4106–0020 (US Government Printing Office, Washington 1971).

PORTER, J. R. and MEYERS, F. P.: Amino acid interrelationships in the nutrition of *Proteus morganii.* Archs Biochem. *8:* 169–176 (1945).

PROSSER, C. L.: Animal models for biochemical research. V. Invertebrates. Introduction. Fed. Proc. Fed. Am. Socs exp. Biol. *32:* 2177–2178 (1973).

PROSSER, C. L. and BROWN, F. A., jr.: Comparative animal physiology; 2nd ed. (Saunders, Philadelphia 1961).

ROE, F. J. C.: Metabolic aspects of food safety (Academic Press, New York 1970).

RUSSELL, P. B.: Antimalarials; in BURGER Medicinal chemistry; 2nd ed., pp. 814–850 (Interscience, New York 1960).

SCHNICK, R. A.: A review of literature on TFM (3-trifluormethyl-4-nitrophenol) as a lamprey larvicide. Invest. Fish Cont. vol. 44, 30 pp. (US Dept. Int. Fisch & Wildlife Service, Washington 1972).

SCHROEDER, H. A.: Recondite toxicity of trace elements; in HAY Essays in toxicology, vol. 4, pp. 107–199 (Academic Press, New York 1973).

SHUGAR, D.: Biochemical aspects of antimetabolites and of drug hydroxylation, vol. 16 (Academic Press, New York 1969).

SPECTOR, W. S. (ed.): Handbook of toxicology. Acute toxicities. vol. 1 (Saunders, Philadelphia 1956).

SMITH, J. N.: Comparative biochemistry of detoxification; in FLORKIN, MASON Comparative biochemistry, vol. 6, pp. 403–458 (Academic Press, New York 1964).

STREET, J. C.: Pesticide interactions. CRC critical reviews in toxicology (CRC Press, Cincinnati 1972).

TOWNSEND, J. F. and LUCKEY, T. D.: Hormoligosis in pharmacology. J. Am. med. Ass. *173:* 44–48 (1960).

VENUGOPAL, B. and LUCKEY, T. D.: Toxicology of non-radioactive heavy metals and their salts. Environ. Qual. Safety *1:* suppl., pp. 4–80 (1975).

WALD, G.: The distribution and evolution of visual systems; in FLORKIN and MASON Comparative biochemistry, vol. 1, pp. 311–345 (Academic Press, New York 1960).

WEISBERGER, A. S. and SUHRLAND, L. G.: The effect of a blocking analogue of cystine on leukemia. J. clin. Invest. *34:* 912–913 (1955).

WILSON, A. and SCHILD, H. A.: Applied pharmacology; 10th ed. (Little, Brown, Boston 1969).

WINTERINGHAM, F. P. W.: Comparative metabolism and toxicology of organic insecticides; in MUNDAY Studies in comparative biochemistry, pp. 107–151 (Pergamon Press, Oxford 1965).

WINTERINGHAM, F. P. W.: Mechanisms of selective insecticidal action. Annu. Rev. Entomol. *14:* 409–442 (1969).

WOOLLEY, D. W.: A study of antimetabolites (J. Wiley & Sons, Chichester 1952).

Prof. T. D. LUCKEY, Department of Biochemistry, University of Missouri Medical School, *Columbia, MO 65201* (USA)

Subject Index

Abalone, as source of food for marine mammals 34
Absorption, effect of food toxicant ingestion on 173
–, of calcium, in marine animals 32, 33
–, of chromium 117
–, of copper 117
–, of minerals, in marine animals 32, 33
– –, in trout 34, 35
–, of selenium 120
–, of vanadium 120
–, of zinc 120
Acantharia 13
Acetocholine esterase, inactivation by organic phosphates 165
Acetylcholine, in protozoa 164
Acheta domesticus 171
Achidoris, magnesium content 31
Achromothrichia, in molybdenum toxicity 115
Achylia gastrica 55
Acid phosphatase, activation by magnesium 45
Acoelomates, toxicology of 168
Aconitase, requirement for iron 21
Acrolein, action in snails 167
Actinomycin D 159
Acyrthosiphon pisum 95
Adaptation, to ingested toxicants 173, 174
Adenyl cyclase, activation by fluorine 12, 18
– –, effect of calcitonin on 39
– – – of calcium on 39
Adrenocorticotrophic hormone, serum zinc level and 121
Aedes aegypti 96
Age, effect of calcium absorption in pig 73
Aging, chromium deficiency during 105
Albatroses, food habits of 33
Alcohol dehydrogenase, requirement for zinc 23
Alcoholic cirrhosis, loss or urinary zinc in 120
Alcoholism, chronic, effect on muscle magnesium 48
Alcyonaria, magnesium content 31
Aldren, effect on insects 173
–, toxicity 163
Algae, iron 30
Alginates 11
Alimentary canal, in nematodes 169
– secretions, copper-binding capacity 118
Alkali disease 116
– metals, excretion of 2
Alkaline phosphatase, activation by magnesium 45
– –, effect of nickel excess on 115
– –, in insects, requirement of zinc for activation 96
– –, in protozoa 94
– –, role in calcium absorption 39
– –, zinc in 23, 121

Allethrin, toxicity 162, 163
Alopecia, in molybdenum toxicity 115
–, in selenium deficiency 111
Aluminium, essentiality of 2
Ameloblasts 38
Amino acids, antagonists of 160
– –, antimetabolites of 153, 154
– –, essential, requirement for 9
– –, free, in insect hemolymph 5
p-Aminobenzoic acid, and sulfanilamide 152
Ammonium phosphate, effect on iron requirement 61
Amphibia, calcitonin in 40
–, trace element requirements of 98, 99
Amylase, activation by chlorine 18
Anabaena cylindrica 15, 17
Anemia, hypochromic 106
– – microcytic, due to iron deficiency 51
–, in copper deficiency 106
–, in molybdenum toxicity 115
–, in selenium toxicity 115
–, microcytic 106
–, normochromic 106
–, normocytic 106
Animal tissues, trace elements in 89–93
Animals, grazing, copper poisoning in 23
– –, molybdenum poisoning in 24
– –, selenium toxicity in 24
–, higher, essentiality of copper in 22
– –, requirement for iron 21
– – – for selenium 23
Annelids, hemoglobin in 51
–, trace element requirements of 96, 97
Anopheles spp., potassium requirement of 16
Antagonism, chemical 152
Antibiotics, action in protozoa 166
–, effect on vinegar eel 169, 170
Anticoagulants, heparins 50
Antimalarials, action in protozoa 166
Antimetabolites 152
–, selected examples of 153
Antimonials, action in platyhelminthes 166
– – in protozoa 166
–, treatment of schistosomiasis with 164
Antimony potassium tartrate, inhibition of phosphofructokinase by 161
Antioxidants, alleviation of selenium deficiency symptoms by 112
Antlers, mineral requirements for 70
Aorta, histological changes, in copper deficiency 106
–, lipid content, effect of magnesium deficiency on 75
–, nickel in 91
Apis fabia 95
– *mellifera* 96
Apposition rate 43
Aquatic animals, mineral requirement of, factors influencing 32, 33
Aragonite, in shells of bivalves 32
Arginase, activation by zinc 23
–, in amphibia 99
–, in fish, dependence on manganese 98
–, in worms 96
Aromatic amines, acetylation of 173
Arsenic, as iodine antagonist 102
–, interaction with iodine 109
–, prevention of selenium toxicity by 116
– trioxide, in rodents 167
Arsenicals, action in protozoa 166
Arthropods, biocides for 167
–, chitin in 10
–, hemoglobin in 51
–, iron requirement of 34
Ascidians, respiratory pigments in 52
Ascorbic acid, effect on iron absorption 53, 78
Aspergillus, calcium deficiency in 17
Aspergillus niger 18
– –, requirement for potassium 16
– – – for sodium 15
– – – for vanadium 19
– – – for zinc 23
Ass, mineral constituents of milk 64
ATPase, role in calcium absorption 39
– – in cellular uptake of calcium 40
–, stimulation by calcium 17
Ataxia, congenital, in manganese deficiency 107
–, neonatal, in copper deficiency 107

Atomic numbers of elements, relation to their abundance 1
Atropine, toxicity of 173
Aurelia 28, 29
Aves, *see* Birds
Avimanganin 20
Azinphosmethyl, toxicity 162
Azotobacter, calcium deficiency in 17
–, requirement for vanadium 19
Azurin 22

Baboons, food habits of 74
Bacillus subtilus 165
Bacteria, calcium deficiency in 17
–, requirement for selenium 23
Bats, food habits of 33
–, piscivorous, food habits of 34
Beduins, osteomalacia in 44
2-4, D (BEE), toxicity 163
Behavior, effect of zinc deficiency on 113
Bephenium-HO-naphthanoate, action in nematodes 166
Bertrand's law 2
Beryllium, essentiality of 2
Bile, copper in 118
Biochemical unity 160, 164
Biocides 166, 167
Biosphere 3
Biotin, sulfur in 10, 19, 48
Birds, bone lesions, due to zinc deficiency 112
–, calcitonin in 40
–, domestic, *see also* Poultry
– –, requirements for calcium and phosphates 58–61
–, growth, manganese requirements for 102
–, insecticide toxicity in 162
–, marine, role of major minerals in 33
–, trace element requirements of 97, 98
–, zinc toxicity in 116
Bivalves, as source of food for marine mammals 33
– – – for trout 36
–, shells, mineral content of 32
Blatella germanica 18, 95
– –, calcium requirement of 17
Blood calcium, regulation of 40, 41
– plasma, iron content of 53
– –, zinc content of, in zinc deficiency 112
– pressure, effect of thiocyanate on 50
–, serum, copper and iron content, in kwashiorkor 107
– –, magnesium content 45
Blowfly, salivary galnd, regulation of secretion by calcium and cyclic AMP 37
Body fluids, ionic composition 28, 55
Bombyx mori 95, 96
Bone and calcium 42–44
–, deformities, in manganese deficiency 108
– density, effect of space travel on 74
–, disorders, in copper deficiency 106
– –, in zinc deficiency 23, 112
–, fluorine in 91
–, formation 37, 38
–, magnesium content of 44
–, marrow, transport of iron in 54
–, metabolism, hormonal regulation 39
–, mineral content, in magnesium deficiency 46
–, osteoporosis, *see* Osteoporosis
–, resorption, in senescence, relation to dietary calcium and phosphorus 41
–, zinc in 92
Bones, of higher animals, presence of calcium and phosphorus 37, 38
–, of sea fish, calcium and phosphorus content 36
Boron, as essential element 2
–, requirement for, in insects 95, 96
–, toxicity of, in insects 95
Brachiopods, hemoerythrin in 52
Brain, demyelination, in copper deficiency 106
–, lesions, in zinc deficiency 113
–, magnesium content 45
–, nickel in 91
Bread flour, fortification with iron 78
Bromide, activation of amylase by 18
–, interchangeability with chloride 18
Bromine, catalytic function 18
Brush border, presence of ATPase in 40
Buffalo, mineral constituents of milk 64

Buoyancy of marine animals 29
1 : 3-Butanediol, as energy source 9

Cadmium, effect on copper absorption 117
– – on zinc requirements, in insects 95
Calcareous tubes, calcium content 30
Calcification, carbonic anhydrase and 31
–, of bone, silicon requirement for 11
Calcite, in shells of bivalves 32
Calcitonin, effect on adenyl cyclase 39
– – on parathyroid secretion 39
–, occurrence in animals 40
–, protection from hypercalcemia by 39
–, role in calcium homeostasis 38
Clacium, absorption, effect of dietary calcium on 40
– – – of phosphate on 40
– –, in intestine 39
– –, in marine animals 32, 33
– –, in subhuman primates 74
– –, in trout 34
– –, inhibition by phytate 41
– and bone 42–44
–, as an antagonist of trace element absorption 4
– carbonate 42
– –, effect on iron requirement 61
– –, in shells of bivalves 32
– –, in skeletons and shells of marine animals 30
–, catalytic function 14, 16
– chelators 159
–, concentration in decapod crustaceans 29
–, content of blood plasma, control of 40, 41
– – of milk of different mammals 64
– – of milk of ewe 66
– – of sea water 33
– – of tissues in mammals 64
– deficiency, induction of osteoporosis by 110
–, difference from the transition elements 31
–, effect on cell permeability 152
– – on iodine metabolism 109
– – on manganese absorption 119
– – – availability 104
– – on parathyroid secretion 39
– – on zinc absorption 120
– – – availability 104
– – – requirement 101
–, essentiality for fish 34
–, excess, effect on iron and zinc absorption in pig 74
– – – on nutrient absorption 151
– excretion, by kidney 40
– –, in decapod crustaceans 29
–, homeostasis, hormonal regulation 38
–, in fish pond culture 34
–, in skeletons and tissues of marine animals 30, 31
–, in urine, of cephalopods 29
–, metabolic role of 16, 17
–, metabolism, disturbances due to high phytate 44
– –, effect of space travel on 74, 75
– –, in higher animals, in relation to requirement 37–44
– phosphate, cycles 37
– – in skeletons and shells of marine animals 31
–, regulation in fish 30
– – in marine animals 28
– – in mollusca 29
– – in tunicates 30
– – of salivary secretion of blowfly by 37
–, relationship to molybdenum, sulfate, zinc and iron 4
–, requirement for antlers 70
– – for egg production and shell thickness 60
– – for maintenance of osmotic and electrolyte balance 6
– – for structural organization 13
– –, of catfish 35
– –, of dog 72
– –, of Drosophila 36
– –, of horses 71, 72
– –, of invertebrates 78
– –, of mammals, factors influencing 64–66
– –, of man 77

– –, of marine animals 32
– –, of pig 73
– –, of poultry 58–61
– – –, for growth 59
– –, of ruminants 66–70
–, role in muscle contraction 38
Calcium-binding protein 39
– –, occurrence 40
Calf, copper poisoning in 113
–, growth, copper requirements for 101
–, magnesium deficiency in 46
–, zinc deficiency in 112
Calliphora erythrocephala 36, 37
Callorhinus ursinus 33
Camel, ash content of milk 64
Canalicula, in bones 37, 38
Carapace of crustacea, requirement of calcium for 32
Carbaryl, toxicity 162
Carbohydrate metabolism, effects of zinc on 121
Carbon, requirement for organic structure 7
– – for photosynthesis 8
– tetrachloride, action in platyhelminthes 166
Carbonates, properties, similarity to phosphates 31
Carbonic acid, formation in marine animals 31
– anhydrase, formation in marine animals 31
– –, zinc in 23, 121
Carboxypeptidase, pancreatic, zinc in 23, 121
Carcharius spp. 36
Cardiovascular disease, effect of hemoglobin reduction on 78
Carnosinase, activation of zinc 23
Carp, artificial feeding of 35
–, food habits of 35
Catalase, iron content of 51
Catalysis, of organic transformations, chemical element needs for 13–24
–, role of macroelements in 14
Catfish, role of calcium and phosphorus in 35
Cations, mobile 14–18
Cattle, cobalt deficiency in 105
–, copper deficiency in 103
–, dairy, fertility of, manganese requirement for 100
–, growth, mineral requirements for 67
–, ill-thrift symptoms in 111
–, manganese deficiency in 108
–, pregnancy, mineral requirements for 67
–, reproduction, manganese requirements for 104
Cebus monkey, magnesium deficiency in 75
Cell division, reole of zinc in 23
– membrane permeability, in protozoa 165
– permeability, balancing by sodium and calcium ions 152
– theory 159
Cellulose 10
–, carbon as a structural component of 9
Cementum, in teeth 38
Central neurotoxins, toxicity 163
Cephalopods, as source of food for marine mammals 33
–, ionic regulation 29, 30
–, skeleton, role of 31
Cerebrospinal fluid, magnesium content 45
Ceruloplasmin 58, 118
–, activity, effect of copper deficiency on 106
– and hemoglobin synthesis 58
–, biosynthesis, effect of copper and zinc on 58
–, copper in 22
–, effect of nickel excess on 115
–, in blood of annelids 96
Cesium, essentiality of 2
Chaenichthyidae 51
Channa punctata 36
Chelating agents, in inorganic nutrition of microbes 93
Chelation of phytic acid to minerals 152
Chemical composition of tissues, constancy in 144
– element needs of animals 1–27
Chick, anemia in copper deficiency 106

–, aorta, changes in copper deficiency 106
–, exudative diathesis, in selenium deficiency 110
–, growth, effect of nickel excess on 115
– – – of selenium deficiency on 102, 111
– –, selenium requirement for 103
– –, silicon requirement for 103
–, growth-promoting properties of molybdenum in 98
–, iron deficiency anemia in 57
–, manganese absorption in 119
– – deficiency in 20, 108
–, requirement for nickel in 22, 98
– – for silicon in 11
– – for vanadium 98
–, selenium deficiency in 24
–, vanadium deficiency in 19
– – toxicity in 116
–, zero equivalent point dose of oxytetracycline in 149
–, zinc deficiency in 113
Child, copper deficiency in 107
–, dental caries in, prevention by fluorine 110
–, zinc deficiency in 23
Chimpanzee, food habits of 74
Chironomids, as source of food for trout 36
Chitin 10
–, carbon as structural component of 9
–, formation, role of phosphorus in 37
Chitonidae 30
Chlamydomonas, sodium requirement of 15
Chloramphenicol 18
–, effect on nematodes 169
Chlordane, action in insects 173
–, toxicity 162
Chloride, concentration in decapod crustaceans 29
–, in lower animals 6
–, interchangeability with bromide 18
–, regulation in fishes 30
– – in marine animals 28
–, requirement for maintenance of osmotic and electrolyte balance 6
Chlorinated hydrocarbons, action in arthropods 167
Chlorine, catalytic function of 14, 18
Chlorocruorin, iron in 30
–, occurrence and role 52
Chlorophyll, magnesium as constituent of 17, 47
Chlorquanide, action in protozoa 166
Chlorsalicylamide, action in platyhelminthes 166
Chlortetracycline 18
Cholecalciferol, role in calcium homeostasis 38
Cholesterol, content of blood, elevation in chromium deficiency 105
Chondroitin mucopoly saccharide, methionine as a structural component of 10
– sulfate, content of cartilage and connective tissue, in manganese deficiency 108
– –, role of 49
– –, synthesis, manganese requirement for 119
– – –, sulfur requirement for 13
–, synthesis of, requirement for manganese 21
Chordates, skeleton role of 31
Chromium, absence from living material 14
–, as growth stimulant 114
–, biologically available forms 103
–, catalytic function 19
–, deficiency, in man 105
– –, in rats and mice 105
–, essentiality of 2
–, metabolism 117
–, occurrence in animal tissues 91
–, requirement for growth 103
– –, of birds 98
– –, of mammals 97
–, toxicity 114
Chronic toxicity 158
Chuppatie-eating Pakistani, phosphorus deficiency in 44
Cilia, in protozoa 165
Circulatory system, in nematodes 169
Citric acid, as chelating agent 93

Citrulline synthesis, activation by fluoride 12
– –, role of fluorine in 18
Cladocera, as source of food for trout 36
Clostridium thermoaceticum, formic dehydrogenase in 24
Cobalt, activation of phosphokinases due to 96
–, as iodine antagonist 102
–, catalytic function 21, 22
– deficiency 19, 21, 22
– –, in cattle, ill-thrift symptoms associated with 111
– –, in ruminants 104, 105
–, interaction with iodine 109
–, occurrence in animal tissues 90
–, requirement, of amphibia 99
– –, of birds 98
– –, of mammals 97
– –, of ruminants, for growth 102
–, toxicity 114
Cocarboxylase, vitamin B_1 as 50
Cod, muscle dipeptidase, dependence of manganese 98
Coelenterates, metabolism 168
–, toxicology of 168, 172
Coenzyme A requirement for sulfur 19
– –, sulfur in 10, 48
Collagen 10, 11
–, formation, role of copper in 106
Collagenase, requirement for calcium 17
Comparative toxicology 161
Competition, for inorganic elements 88
Competitive inhibitors 152
Confused fluor beetle, trace element requirements of 95
Congenital abnormalities, in zinc deficiency 113
Connective tissue, silicon in 120
Cooking, effect on essential nutrients 154
Copepods, carbon and nitrogen uptake 34
Copper absorption and utilization, effect of zinc on 116
– –, defect in 106
– –, effect of dietary components on 117
–, catalytic function of 22
–, competition with other minerals 150
–, complexes in feed, availability of 117
–, content of blood, in kwashiorkor 107
– –, of liver, effect of nickel excess on 115
– – –, in protein-caloric malnutrition 107
– deficiency 3, 105, 106
– –, in dairy cattle 103
– –, in livestock 22
– –, in poultry 98
– –, induction by molybdenum 115
– –, overcoming by selenium 111
–, effect on calcium storage 33
– – on sulfur utilization in ruminants 71
–, essentiality of 2
–, excess, effect on growth of pigs 148
–, in hemocyanin 52
–, interaction with zinc 95
–, interrelationship with molybdenum and iron 118
– – with other elements 23
–, occurrence in animal tissues 90
–, poisoning 24, 113
–, protective effect against molybdenum toxicity 115
–, relationship to molybdenum and sulfate 24
–, requirement for gestation 99
– – for growth 101
– – for reproduction 103
– –, of amphibia 99
– –, of fish 98
– –, of honeybee 95, 96
– –, of insects 95, 96
– –, of protozoa 94
– –, of worms 96, 97
–, role in iron metabolism 58
– – in phospholipid synthesis 118
– – in thyroid hormone synthesis 109
–, storage in liver 90
–, toxicity 22
–, utilization, in ruminants, effect of molybdate and sulfate on 70
– –, interference by zinc excess 58
Copper-binding protein 117
Corals, iodine in 18
–, phosphates in 13

–, requirement for calcium 32, 78
–, skeleton, calcium content 30
Coumarin, action in rodents 167
Cow, lactation, mineral requirements for 68
–, reproduction, role of manganese in 109
–, safe levels of fluoride 114
–, selenium deficiency 103
–, tolerance to high copper intake 113
Crabs, as source of food for marine mammals 33
Cricket, activity spectrum of insecticides in 171
–, effect of pentoses on 151
–, growth of, effect of sodium on 146
–, pentose toxicity in 151
Crithidia fasciculata 94
Crustaceans, artificial feedings of 35
–, as source of food for marine birds 33
–, calcium in 13
–, insecticide toxicity in 162
–, magnesium content 31
–, requirement for calcium 32, 78
–, skeleton, role of 31
Ctenophora, buoyancy of 29
–, metabolism 168
–, toxicity of 168
Culex pipiens 96
Culture media, for protozoa 93
Cyanides, conversion to thiocyanates 50
Cyanocobalamin, requirement of birds 98
Cyclic AMP and parathyroid hormone secretion 39
– –, regulation of salivary secretion in blowfly by 37
Cyclostomes, sekeleton, role of 31
Cystine, as derivative of methionine 48
–, as source of sulfur 48
Cytochrome b5 24
– c 24
–, iron in 30
– oxidase 118
– –, activity, effect of nickel excess on 115
– – –, in copper deficiency 107
– – –, of leukocytes, in iodine deficiency 109
– –, copper in 22
– –, in protozoa 94
– –, requirement for iron 21
Cytochromes, iron content of 51
–, requirement for iron 21
–, role in electron transport 52

Dalapon, toxicity 163
DDT, action in arthropods 167
–, adaptation of insects to 173
–, toxicity 162
Decapod crustaceans, ionic composition 29
– –, sekeleton, calcium content of 31
Decarboxylases, role of manganese in 20
Decarboxylation, magnesium as cofactor in 45
Deficiency, of inorganic elements, conditioned 89
– – –, simple 89
–, of trace elements 104–117
Deldrin, toxicity 163
Dental caries, effect of fluoride on 97
– –, prevention, by fluorine 12, 110
– – – by vanadium 19
Dentin, in teeth 38
Detoxication 164
Detoxification, mechanisms in insects 173
Development, post-natal, trace element requirements for 100–103
– –, zinc requirement for 101
–, pre-natal, trace element requirements for 99, 100
Di-amine oxidase, association of copper with 118
Diarrhoea, chronic, magnesium deficiency in 48
Dibromindigo 18
Dibromotyrosine 18
Dichlobenil, toxicity 163
Dichloronitrosalicyl amilide 171
Dichlorophen, action in platyheleminthes 166
Dieldrin, action in insects 173
Diethylcarbamazine, action in nematodes 166
Diets, for insects 78, 95

–, for monkeys 75
Differentiation, trace element requirements for 99, 100
1,25-Dihydroxycholecalciferol, role of 39
Diiodotyrosine 18
–, synthesis, effect of goitrogens on 50
9,10-Dimethyl-1,2-benzanthracene 113
Dipeplidase, in fish, dependence on manganese 98
Diploblasts, toxicology of 168
Diplocardia longa 96
Diquat, toxicity 163
Diversity 160
–, in comparative nutrition 144
Dizinon, toxicity 162
Djenkolic acid, sulfur in 48
DNA, function, role of zinc in 23
– polymerase, zinc in 121
–, stabilization by nickel 22
–, synthesis, dependence on zinc 121
Dog, anemia in copper deficiency 106
–, mineral constituents of milk 64
– – requirements of 72
–, osteoporosis in 110
Dolphin, food habits of 33
Drosophila melanogaster 96
Drosophila, requirement for calcium in 36
Duck, effect of copper excess in 114
–, selenium deficiency in 111

Eared seal, food habits of 33
Echinarachnium parma 13
Echinodermata, magnesium content 31
Echinoderms, regulation of ions in 29
Eel, artificial feeding of 35
Egg, hatchability, effect of selenium deficiency on 112
–, production, calcium requirements for 60
– –, effect of iodine excess on 114
– – – of selenium toxicity on 116
– –, manganese requirements for 100
– –, phosphorus requirements for 60
–, shell-thickness, calcium and phosphorus requirement for 60
– –, in manganese deficiency 108
–, zinc in 93
Eggs, of marine animals, calcium and magnesium content 31
Elasmobranchs 5
Elastin 10
– biosynthesis, defect in, in copper deficiency 106
Electrolyte balance, in animals 5–7
Electron transport, iron compounds facilitating 52
Elements, essentiality of, criteria for establishing 4
Elephant seals, food habits of 33
Embryonic mortality, effect of iodine excess on 114
– – – of selenium deficiency on 111
Enamel, in teeth 38
Endopterygotes, chitin formation in 37
Endothelial degeneration, in selenium deficiency 112
Endrin, toxicity 163
Energy, sources, for ruminants 21
– –, unconventional and synthetic, in animal nutrition 9
–, utilization, toxicants affecting 159
Energy-rich phosphates, synthesis, role of calcium in 38
Entamoeba histolytica 167
Entercoela, toxicology of 172, 173
Enzootic marasmus, in cobalt deficiency 104
Enzymes, activation by magnesium 45
–, calcium-sensitive 38
–, copper-containing 118
–, inactivation by organic phosphates 165
–, induction, role in detoxification of food toxicants 173
–, iron sulfur flavin 49
–, iron-containing 52
–, manganese-containing 119
–, molybdenum-containing 119
–, proteolytic, action in nematodes 167
–, vitamin B_{12}-containing 21
–, zinc-containing 23, 121
Eobiont 14
Ergothionine, sulfur in 48
Erythrocytes, hemoglobin in 52
–, production, in copper deficiency 106

Erythropoiesis 56
Escherichia coli, formic dehydrogenase in 24
Esophagus lesions, in zinc deficiency 112
Essential nutrients, toxicity of 151–152
Euglena gracilis 94
Euhydra lutris 34
Eulex molestus 96
Eupargurus 28
Evolution 144
Ewe, effect of iodine excess on 114
–, lactating, mineral requirements of 69
–, pregnant, mineral requirements of 68
Excretion, of cations vs. anions 2
–, of minerals, in mammals 66
Excretory system, in nematodes 169
Exudative diathesis, due to selenium deficiency 24, 110, 111
Eye, lesions, in chromium deficiency 105

Fat excess, effect on nutrient absorption 151
Feathering, effect of zinc deficiency on 112, 113
Fenac, toxicity 163
Fenitrothion, toxicity 162
Ferredoxin, in protozoa 167
Ferric ammonium citrate, as source of iron 56
– orthophosphate, as source of iron 56
Ferritin 54
–, as catalyst in sulfide oxidation 51
–, as storage form of iron 54
–, iron content of 51
Ferrous ascorbate, absorption 56
– sulfate, bioavailability 55
Ferroxidase I 118
–, activity, in copper deficiency 106
– –, of ceruloplasmin 58
Fertility, effect of fluoride deficiency on 97
– – of manganese deficiency on 109
– – of selenium deficiency on 110, 111
– – – toxicity on 116
–, of dairy cattle, manganese requirement for 100
–, of mouse, effect of fluorine deficiency on 110
Fertilizers, inorganic, for fish pond cultures 34
Fetus, chromium content 117
–, resorption, in copper deficiency 106
Fish, as source of food for marine birds 33
– – – – mammals 33
–, ionic regulation in 30
–, iron requirement of 34
–, mineral requirements of 34–36
–, pond culture, inorganic fertilizer in 34
–, skeleton, role of 31
–, trace element requirements of 98
Fishes, warm water, artificial feeding of 35
Flagellates, effect of trichomonicides on 167
Fluid medium, chemical element needs for 5–7
Fluoride, as iodine antagonist 102
–, effect on iodine metabolism 109
–, silicon as antagonist of 12
Fluorine, catalytic function 15, 18
–, content of water, effect on fluorosis 110
– deficiency 109, 110
– –, effect on reproduction 110
–, essentiality of 12
–, in food, effect on hard tissues 110
–, metabolism 118
–, occurrence in animal tissues 91
–, requirement of mammals 97
–, role in prevention of dental caries 109, 110
–, toxicity 114
Fluoroacetate, as antimetabolite 152
Fluorofatty acids 18
Fluorosis, relationship to fluorine content of water 110
Folic acid, in liver, effect of zinc on 121
Food additives 154
– –, intentional 158
–, maternal, definition of 148
– processing, effect on nutrients 154
– toxicants 154–158
Foraminifera, magnesium content 31
Formic dehydrogenase, role of selenium in 24
Fox, eye, zinc in 92

Fracture healing, role of chondroitin sulfates in 49
Fructokinase, potassium requirement for 16
Fulmars, food habits of 33
Fungi, calcium deficiency 17

Galactose oxidase, presence of copper in 22
Gasterophilus 6
Gastrointestinal tract, disorders, in copper deficiency 106
– –, losses of iron from 76
Gastroliths 13
Gastropods, bromine in 18
–, ionic regulation 29
–, iron content 30
–, magnesium content 31
Germ-free chicks, growth of, effect of oxytetracyclene on 147
– rats, zinc requirement of 121
Gestation, copper requirement for 99
–, of rat, manganese requirements for 100
Gills, absorption of minerals through 34
–, of decapods, ion absorption by, against gradient 29
–, of lobster, impermeability to magnesium and sulfate ions in 29
–, of teleosts, excretion of ions through 30
Glucagon, regulation of blood glucose by 40
Glucose, metabolism, in chromium deficiency 105
– tolerance factor 20, 97, 117
Glutamic dehadrogenase, in amphibia 99
– –, zinc in 23
Glutathione peroxidase, selenium in 24, 120
–, sulfur in 48
Glycerol, content, of insects 5
α-Glycerophosphate, oxidation, effect of nickel deficiency on 22
Glycine dehydrogenase, role of selenium in 24
Glycogen, of leukocytes, in iodine deficiency 109
Glycolysis, requirement of potassium for 16
–, role of calcium-sensitive enzymes in 38
Glycosaminoglycans, silicon in 11
Glycosyltransferases, activation by manganese 21
Goat, copper deficiency in 107
–, manganese deficiency in 108
–, mineral constituents of milk 64
–, reproduction, role of manganese in 109
Goitre, induction by arsenic 109
–, relationship to iodine intake 102
Goitrogenic effects, of thiocyanate 50
Goitrogens, interference with thyroid hormone synthesis 19
–, occurrence 50
–, sulfur in 50
Golden hamster, calcium excretion in 40
Gorgonin, bromine in 18
Gorilla, food habits of 74
Growth, effect of nickel toxicity on 115
– – of selenium deficiency on 110
– – of zinc deficiency on 112
– hormone, effect on synthesis of chondroitin sulfates 49
–, of chicks, effect of selenium deficiency on 111
–, of insects, effect of trace element deficiency on 95
–, of mammals, accretion of calcium during 64
– –, deposition of calcium during 65
–, of mouse, effect of fluorine deficiency on 110
–, of rats, stimulation by chromium 114
–, requirement for manganese 102
– – for molybdenum 103
– – for selenium 102, 103
– – for silicon 103
– – for tin 103
– – for trace elements 100–103
– – for vanadium 103
– – for zinc 101
Guinea pig, copper deficiency in 107
– – – requirement of 100
– –, manganese deficiency in 108
– –, mineral requirements of 72
– –, reproduction of, role of manganese in 109

Haemerythrin, presence of iron in 30
Haemochromogen, presence of iron in 30
Hair, depigmentation, in copper deficiency 106
–, growth, effect of zinc deficiency on 112
–, loss, in selenium toxicity 115
–, zinc in 92
Halogens, catalytic function 18, 19
Hamster, effect of zinc on tumor development in 113
–, mineral requirements of 72
HCN 159
–, toxicity of 173
Heart failure, in copper deficiency 106
Heavy metals, catalytic function 19–24
– –, toxicity 159
Heme, as prosthetic group of chlorocruorin 52
– – – of hemoglobin 52
–, in hemoglobin 52
–, systems, requirement for iron 21
Hemichordates, hemoglobin in 51
Hemochromatosis, effect on iron absorption 53
Hemocyanin, occurrence 52
– oxygen carriers, of invertebrates 22
Hemoerythrin, occurrence and role 52
Hemoglobin, absorption, effect of phytate on 53
–, biosynthesis, and ferroxidase activity of ceruloplasmin 58
–, content of, blood plasma 53–54
–, in acoelomates 168
–, in nematodes 169
–, in protozoa 164
–, iron content of 30, 51
–, occurrence and role of 51–52
–, reduction, effect on survival from cardiovascular disease 78
– repletion test 57
–, requirement for iron 21
–, synthesis, requirement for iron 78
Hemoproteins 49
Hemosiderin, as storage form of iron 54
–, iron content of 51
Hemosycotypin, zinc in 23
Hemovanadin, occurrence and role 52
Hen, laying, effect of iodine excess on 114
Heparin, maintenance of blood fluidity by 49
–, occurrence 50
–, sulfur content 50
–, synthesis, requirement for manganese 21
Hepatosis dietetica, in selenium deficiency 24, 110
Hepatotoxins, toxicity 163
Heptachlor, action in insects 173
–, toxicity 163
Herbicides, toxicity 163
Hexaresorcinol, action in nematodes 166
Hibernation and serum magnesium 46
Hilsa ilsha 36
Holothurians, hemoglobin in 51
Homarus vulgaris 93
Homerus 28
Homocystine, derived from methionine 48
Honeybee, requirement for trace elements 95
Hoppocampal region of brain, development, role of zinc in 121
Hormetic, definition 148
Hormesis, definition 148
Hormology, definition 148, 149
–, of essential nutrients 147
Hormones, antimetabolites of 153
Horse, mineral constituents of milk 64
–, requirement for major minerals, factors affecting 71, 72
Hyaluronic acid synthesis, role of manganese in 21
Hydration of tissues, role of chondroitin sulfates in 49
Hydrogen, catalytic function of 14
– ion concentration, of internal medium, of different animals 7
– requirement, for organic structure 7
– sulfide, metabolism in mammals 51
– transport, blocking by trichomonicides 167
Hydrolases, role of manganese in 20
Hydrolytic enzymes, requirement for magnesium 17

Hydroperoxidase, requirement for iron 21
Hydroxyapatite 13, 37, 42
–, fluorine in 18
Hydroxylase, role in formation of 1,25-dihydrocholecalciferol 39
Hypercalcemia, prevention by calcitonin 39
Hyperemia, of liver, molybdenum toxicity due to 109
Hyperparathyroidism, due to phosphorus excess 42
Hypogonadism, effect of zinc deficiency 112

Ictolurus punctatus 35
Idiopathic hypercalcemia 44
Ill-thrift syndrome, due to selenium deficiency 102, 111
Imbalance, between essential nutrients 151
Infant, copper deficiency in 107
–, malnutrition, role of magnesium in 47
–, premature, copper deficiency in 107
Infection, role of selenium in overcoming 111
Ingested material, fate of 145
Inhibition, competitive 153
–, noncompetitive 153
Inorganic elements, biochemical significance 88
– –, conditioned deficiency 89
– –, essentiality of 87
– –, requirement, functional 88
– – –, physiological 88
– – simple deficiency 89
Inositol hexaphosphate, chelation to minerals 152
Insect adaptation, to insecticides 173
Insecticides, activity spectrum, in crickets 171
–, adaptation of insects to 173
–, comparative toxicity 162, 163
–, effect on schizocoela 170
–, selective toxicity of 165
Insects, as source of food for trout 36
–, experimental, diets for 78
–, insecticide toxicity in 162
–, mineral requirements of 36, 37
–, phytophagous, requirement for potassium and magnesium 6
–, potassium requirement of 16
–, resistance to insecticides 173
–, trace elements, requirement of 94–96
–, zoophagous, sodium requirement of 6
Insulin, chromium as cofactor for 20
–, regulation of blood glucose by 40
Integment, of lobster, impermeability of magnesium and sulfate ions in 29
Intestinal tract, protozoa in 165
Intestine, calcium absorption 39
Invertebrates, calcium requirement of 78
–, copper toxicity in 22
–, essentiality of copper in 22
–, hemoglobin in 52
–, iron requirement of 21
–, marine, osmotic pressure of external media in 28
–, respiratory pigments in 51, 52
–, stimulation by iodoamino acids 19
–, trace element requirements, insects 94–96
– – – –, protozoa 93, 94
– – – –, worms 96, 97
–, vanadium in 19
–, zinc requirement of 23
Iodine, as essential element 2
– – –, early studies on 1
–, catalytic function 18, 19
–, counteracting goitrogenic potential of thiocyanate by 50
– deficiency 109
– excess, tolerance of various species to 114
–, interactions with other elements 109
–, metabolism 118
–, occurrence in animal tissues 91
–, requirement for growth 102
– – for, in amphibia 99
– – for, in mammals 97
Iodonin 18
Iranians, disturbances of zinc and calcium metabolism due to high phytate 44
Iron, absorption 53
– –, defects in 106

– –, effect of manganese on 115
– – – of zinc on 116
– –, in monkey, effect of soy protein on 75
– –, in pig, effect of minerals on 74
–, as essential element, early studies on 1
–, availability, in chicks and rats 57
– balance, regulation of 56
–, catalytic function 21
–, competition with other minerals 150
–, content of animal tissues 30
– – of blood 51, 52
– – –, in kwashiorkor 107
– – of milk on different mammals 64
– deficiency anemia 57
– – –, effect of vitamin C on 78
– – –, in chickens 57
– – –, susceptibility to manganese toxicity in 115
– –, effect on iron absorption 53
– –, in man 51
–, enrichment, of wheat flour 55
–, essentiality for fish 34
– – of 2
–, excretion 54
– fortification, policy in UK 78
–, intake, protection against copper poisoning by 113
–, interrelationship with copper 23
– – – and molybdenum 118
–, losses in man 76
– metabolism 53, 54
– –, effect of manganese deficiency on 114
– –, in relation to requirements 51–58
– –, role of copper in 58
– – – of vanadium in 19
– mobilisation, from liver stores, role of ceruloplasmin in 118
–, requirements 54–58
– –, of horses 71, 72
– –, of insects 95, 96
– –, of man 77
– –, of pig 73
– –, of poultry 61
– –, of subhuman primates 75
–, role in catalysis 14
–, storage 54
– – components 51
– sulfur flavin enzymes 49
– transport 53, 54
– –, in green plants, role of silicon 12
–, utilization, inhibition by phosphate 41
– – of, effect of zinc on 116
Iron-binding capacity, in serum, in kwashiorkor 107
Iron-sulfur proteins, metabolism of 49
Isocitric dehydrogenase activity, effect of nickel excess on 115
– –, in protozoa 94
Isoleucine, competition with valine and leucine 152
Isothiocyanate, esters, conversion to thiocyanate 50
– sulfur in 48

Japanese quail, growth, effect of selenium deficiency on 102
Jelly fish, as source of food for marine birds 33

Keratin, methionine as structural component of 10
Keratogenesis defects, in zinc deficiency 112
Ketoacids, decarboxylation, role of vitamin B_1 in 50
Kidney, magnesium content of 45
–, role in regulation of serum calcium 40
–, selenium in 92
Killer whale, food habits of 33
Kinases, role of manganese in 20
Kittiwakes, food habits of 33
Krill, as source of food for marine mammals 33
Kwashiorkor, effect on trace element concentration of blood 107
–, magnesium deficiency in 48

Laboratory animals, mineral requirements of 72, 73
Lactate, content, in *Gasterophilus* larvae 5
Lactation, calcium excretion in 66
–, in ruminants, requirement of major mineral elements for 69

–, trace element requirements for 103, 104
Lactic dehydrogenase, in protozoa 94
– –, zinc in 23
Lactobacillus arabinosis 20
Lactose synthesis, requirement for manganese 21
Lacunae 43
–, in bones 37, 38
Lamb, copper deficiency in 107
– – – in, anemia in 106
Lamb, growth, molybdenum requirement for 103
– – of, effect of selenium deficiency on 102
–, manganese toxicity in 115
–, molybdenum deficiency in 97
–, selenium deficiency in 111
–, zinc deficiency in 23
– – toxicity in 116
Lamellibranchs, hemoglobin in 52
–, ionic regulation in 29
–, myoglobin in 52
Lamprey, toxicology of 172
Lampreycide, comparative toxicity of 172
Laying hen, calcium requirements of 60
– –, magnesium requirement of 61
– –, phosphorous requirement of 60
Leg deformities, in zinc deficiency 113
Leopard seal, food habits of 33
Leptocephalus 51
Lethargy, in selenium toxicity 115
Leucine, competition with isoleucine and valine 152
Leukocytes, cytochrome oxidase activity, in iodine deficiency 109
–, phagocytic and bactericidal properties, in iodine deficiency 109
Libido, effect of manganese deficiency on 109
Ligands 14
Ligia oceanica 28
Limpets, iron in 30
Lindane, action in insects 173
–, toxicity 163
Lineus 30
Lingula 52
Lining, in fish pond culture 34
Lipase, requirement for calcium 17
Lipemia, in iron deficiency anemia 57
Lipid metabolism, role of vanadium in 19
Lipoic acid, sulfur in 10, 19, 48
Lithium, essentiality of 2
Lithosphere 3
Liver, copper stored in 90
–, damage, in selenium toxicity 115
–, degeneration, in chicks, in selenium deficiency 110
–, fatty degeneration, in tin toxicity 116
–, magnesium content of 44, 45
–, neonatal, copper content 101
Lobodon carcinophagus 33
Lobster, ionic composition 29
–, zinc in 93
Locust, mineral requirements of 36
Longevity, effect of tin toxicity on 116
–, of rats and mice, effect of chromium on 105
Lucanthene, action in platyhelminthes 166
Luciferase, in worms 96
Lymantia dispar 96
Lysine oxidase, copper in 22
Lysol oxidase 118

M. persicae 95
Macroelements, catalytic function of 14
Magelona 52
Magnesium carbonate, availability 46
–, catalytic role of 14, 17, 18
–, content of animal tissues 45
– – of average diet 47
– – of milk of different mammals 64
– – of sea water 33
– – of tissues and skeletons of marine animals 31
– deficiency, alleviation by fluorine 110
– –, effect on intracellular concentration of magnesium 45
– –, in children 47, 48
– –, in monkeys 75
– –, in rats 46
–, effect on calcium absorption 33
–, essentiality for fish 34

–, excess, effect on iron and zinc absorption in pig 74
– excretion 47, 48
– – in decapod crustaceans 29
– in urine, in cephalopods 29
–, metabolism, in relation to requirements 44–48
–, occurrence 47
–, regulation in echinodermes 29
– – in fishes 30
– – in marine animals 28
– – in tunicates 30
–, requirement for antlers 70
– – for growth of poultry 61
– – for maintenance of osmotic and electrolyte balance 6
– –, of dog 72
– –, of horses 71, 72
– –, of man 77
– –, of pig 73
– –, of ruminants 66–70
– –, of subhuman primates 75
–, salts, absorption 47
– –, availability 46
–, utilization of, inhibition by phosphate 41
Malaoxon, formation from malathion 164
Malathion, conversion to malaoxon 164
–, toxicity 162
Malic dehydrogenase, activity, effect of nickel excess on 115
– –, in protozoa 94
– –, zinc in 23
Malnutrition, copper deficiency in 107
–, *in utero* 100, 101
Maltase, requirement for calcium 17
Mammals, biocides for 167
–, insecticide toxicity in 162
–, marine, role of major minerals in 33, 34
–, mineral constituents of milk 64
–, requirement for major mineral elements, factors affecting 62–66
–, trace element requirement of 97
–, zinc toxicity in 116
Man, calcitonin in 40
–, chromium deficiency in 19, 105
–, cobalt deficiency in 22
–, copper absorption, defect in 106
– – deficiency in 22, 107
–, dental caries in, prevention by fluorine 110
–, effect of cobalt excess in 114
–, iodine requirements for growth 102
–, iron absorption in 55
– – deficiency in 51
– – requirements of 54, 55
–, magnesium excretion in 47
– – requirement of 47
– – toxicity in 115
–, mineral constituents of milk 64
–, phosphorus deficiency in 41, 44
–, requirement for major mineral elements 75–78
–, tin metabolism in 119
–, zinc deficiency in 23, 112
– – metabolism in 120
Manganese, activation of arginase by 96
– – of phosphokinases by 96
–, as iodine antagonist 102
–, availability, effect of calcium on 104
–, catalytic function of 20, 21
–, competition with other minerals 150
–, content of animal tissues 91
– – of liver, in protein-calorie malnutrition 107
– – of thyroid gland, in patients suffering from goiter 109
– deficiency 3, 19, 20, 107–109
– –, in poultry 98
–, excess, effect on iron and zinc absorption in pig 74
–, metabolism 118, 119
–, requirement for gestation 100
– – for reproduction 104
– –, of amphibia 99
– –, of birds 97, 102
– –, of fish 98
– –, of honey-bee 95, 96
– –, of insects 95, 96
– –, of mammals 97
– –, of poultry 104
– –, of protozoa 94
– –, of worms 97

–, role in thyroid hormone synthesis 109
–, toxicity, interference with iron metabolism 114
Marine animals, requirement for major mineral elements in 78
– –, role of major minerals in 28–32
– –, skeleton, major minerals in 31, 32
– –, source of mineral elements for 78
– –, tissues and skeleton, major minerals 30–32
– birds, mineral elements in 33
– –, role of major minerals in 33
– decapods, ionic composition 29
– elasmobranchs, ionic regulation in 30
– mammals, mineral elements in 33
– teleosts, requirement for minerals 33
Mating, effect of manganese deficiency on 109
Membrane permeability, comparative 161
– –, toxicants affecting 159
Menkes' kinky hair syndrome 106
Menstruation, losses of iron during 76
Mesoderm, in nematodes 169
Metabolic acids, in *Gasterophilus* larvae 6
– rate, in magnesium deficiency 47
Metabolite-receptor complex 152
Metabolizable energy content of diet, nutritional requirements based on 60
Metal-enzyme complexes 15
– –, role of manganese in 20
Metalloenzymes 15
–, distinction from metal-enzyme complexes 89
Metalloproteins, manganese in 20
Metamorphosis, nutrient requirements for 99
–, of blowfly, metabolism of phosphorus in 37
Metazoa 167
Methionine, as source of sulfur 48
–, oxidation of 48
–, requirement 10
Methoxychlor, resistence of insects to 173
–, toxicity 162
Methyl groups transfer, role of vitamin B_{12} coenzymes in 22
Methylmalonyl-CoA isomerase 21
Metronidazole 167
Microorganisms, inorganic element requirements of, effect of temperature on 94
Milk fish, artificial feeding of 35
–, magnesium content 45
–, mineral constituents of different mammals, table 64
Mineral elements, excretion, in mammals 66
– –, major 28–86
– – –, in tissues and skeleton, of marine animals 30–32
– – –, of aquatic animals 32, 33
– – –, requirements of mammals, facotrs affecting 62–66
Mineralization, of bone, silicon requirement for 11
Mink, iron deficiency in 34
Mirex, action in arthropods 167
Mironuga leonina 33
Mitochondria, calcium transport 39
–, in protozoa 165
–, oxidative phosphorylation in 52
–, ultrastructural changes in vitamin D deficiency 39
Molluscicides 171
Molluscs, as source of food for marine birds 33
– – – for trout 36
–, calcium in 13
– – requirement of 32, 78
–, ionic composition of 29
–, iron requirement of 34
–, shells, calcium content 30
–, skeleton, role of 31
Molybdate, effect on copper utilization in ruminants 70
– – on sulfate reduction by rumen microorganisms 70
Molybdenum, absorption 119
–, as essential element 2
–, catalytic function of 24
– deficiency 24
–, effect on copper absorption 117
– – – requirement 101

–, excess, tolerance of various species to 115
–, growth-promoting properties, in chicks and turkey poults 98
–, interrelationship with copper and iron 118
– – – and sulfate 23
–, occurrence in animal tissues 91
–, poisoning 24
–, requirements, of amphibia 99
– –, of chicks, effect of tungsten on 98
– –, of insects 95, 96
– –, of lambs, for growth 103
– –, of mammals 97
– –, of protozoa 94
– –, of worms 96, 97
–, toxic effects, abolishment by iodine 109
–, toxicity 115
– –, alleviation by sulfate 119
– – of, in insects 95
Monkeys, requirements for major minerals 74, 75
Monoamine oxidase, association of copper with 118
– –, in amphibia 99
Mosquito, abdomen, zinc content 96
Mouse, fluoride deficiency in 97
–, growth, effect of nickel toxicity on 115
– –, manganese requirement for 102
–, manganese deficiency in 108
–, mineral requirements of 72
–, reproduction, effect of fluorine deficiency on 110
–, tin toxicity in 116
Mucoitin sulfate, occurrence 49
– –, role of 49
Mucopolysaccharide synthesis, requirement for manganese 20
Muscle contraction, role of calcium in 38
– hemoglobin, *see* Myoglobin
–, magnesium content of 44
– necrosis, in selenium deficiency 111
Muscular dystrophy, due to selenium deficiency 24
Mutation, as protective mechanism 173
Myoglobin, iron content of 51
–, occurrence in animals 52
–, requirement for iron 21
Myopathy, in lambs 116
–, in selenium deficiency 110
Myostracal layers, aragonite in 32
Mytilacea 32
Mytilus californianus 32
Myxinoids, ionic regulation in 30
Myzus perscae 95

NAD-linked dehydrogenases, role in electron transport 52
NAS Committee on Water Quality Criteria 165
Natilba 30
Necrosis, of muscle, in selenium deficiency 111
Nekton, as source of food for marine mammals 33
Nematodes, biocides for 166–167
–, toxicology of 169
Nemertines, hemoglobin in 51
–, iron content 30
–, myoglobin in 52
Neomyzus circumflexus 95
Nephrosis, loss of urinary zinc in 120
Nephthys 30
Nervous system, central, effect of zinc deficiency on 113
Neutropenia, in copper deficiency 107
New Zealand, selenium deficiency in grazing ruminants 111
Nickel, absence from living material 14
–, absorption 119
–, catalytic function of 22
–, competition with other minerals 150
–, essentiality of 97
–, occurrence in animal tissues 91
–, requirement of chicks 98
–, toxicity 115
Niclosamide, action in platyhelminthes 166
Nitrate reductase, role of molybdenum in 24
Nitrilotriacetic acid, as chelating agent 93
Nitrogen cycle, in sea, role of plankton in 34

– fixation, requirement for vanadium 19
– –, role of cobalt in 22
– – – of iron in 21
– requirement 9
– – for organic structure 7
Nitrogenase, requirement for iron 21
Nitrosomonas, calcium deficiency in 17
Noctilio, food habits of 34
Noosphere 4
Norbormide, susceptibility of rats to 173
Nucleic acid metabolism, role of zinc in 23
Nurture, definition 144
Nutrition, definition 144

Ochromonas malhamensis 94
Octocalcium phosphate 42
Odobenus, food habits of 33
Odontoceti 33
Oestrus, cycle, effect of copper on 103
– – – of zinc on 104
–, defects, in selenium deficiency 111
–, depression, in manganese deficiency 109
Oncopeltus fasciatus 96
Orca, food habits of 33
Organophosphates, action in arthropods 167
Osmotic balance, mineral requirement for 5
– regulation, in marine animals 78
– –, in nematodes 169
Ossification, role of chondroitin sulfates in 49
Osteoblasts, in bones 37, 38
Osteocytes, in bones 37, 38
Osteocytic osteolysis 43
Osteoids, in bones 37, 38
Osteomalacia, etiology of 44
Osteoporosis, due to impairment of vitamin D metabolism 76
–, due to phosphorus excess 42
–, effect of fluorine on 110
– – of iron deficiency on 57
Otolith development, in manganese deficiency 108
Ovulation defects, in selenium deficiency 111
Ox, mineral constituents of milk 64
Oxalacetic decarboxylase, activation by zinc 23
Oxidative phosphorylation 52
– –, magnesium as cofactor for 17, 45
Oxygen, establishment upon earth 7
– requirement, for organic structure 7
Oxytetracyclene, effect on growth rate of chicks 147
–, zero equivalent point dose in chicks 147, 149

Pakistani, phosphorus deficiency in 44
Pancreas, exocrine, degeneration, in selenium deficiency 111
Parakeratosis, in zinc deficiency 112
Parathion toxicity 162
Parathyroid gland, calcitonin in 40
– hormone, interdependence of vitamin D and 39
– –, role in calcium homeostasis 38
– – secretion, stimulation by calcitonin 39
Parathyroids, regulation of blood calcium by 40
Parturition, effect of zinc deficiency on 112
Patella vulgata 30
Patellidae 30
Pectin 11, 17
Penicillium, calcium deficiency in 17
Penguin, as source of food for marine mammals 33
–, food habits of 33
Pentoses, effect on growth of crickets 150
–, toxicity, in crickets 151
Peptidases, activation by zinc 23
–, requirement for magnesium 45
Periodontal disease, in selenium deficiency 24
Pernicious anemia, due to cobalt deficiency 22
Perosis, in chick, in manganese deficiency 108
Peroxidase activity, of leukocytes, in iodine deficiency 109
Peroxidases, dependence on iron 52
Pesticides, in food 154, 158

Petrels, food habits of 33
Phagocytosis, of leukocytes, in iodine deficiency 109
Phalaris staggers, effect of cobalt on 105
Phalaris tuberosa, cobalt deficiency due to 105
Phallusia 28
Pheasant, zinc deficiency in 113
Phenols, conjugation of, source of sulfate for 51
Pheromones 158
Phoronids, hemoglobin in 51
Phosphatase, activation by magnesium 45
–, requirement for calcium 17
Phosphate, absorption, in marine animals 32, 33
– –, relationship to calcium transport 40
–, catalytic function of 14
–, effect on iron absorption 53
– – on manganese absorption 119
– – on nutrient absorption 151
–, excess, inhibition of utilization of other minerals by 41
–, in skeleton, of marine animals 31
–, organic, inactivation of enzymes by 165
–, properties, similarity to carbonates 31
–, requirement for maintenance of osmotic and electrolyte balance 6
–, role in calcium absorption 40
Phosphofructokinase, inhibition of 161
Phosphokinases, in worms, activation by copper or manganese 96
Phospholipid synthesis, role of copper in 118
Phosphorus, absorption, in trout 35
–, content of milk of different mammals 64
–, cycles, in sea, role of plankton in 34
– deficiency, in catfish 35
– –, in man 41, 44
– –, in turkey 60
–, excess, senile osteoporosis due to 42
–, in fish pond culture 34
–, metabolism, effect of space travel on 74, 75
– –, in higher animals 37, 38
– –, in metamorphosis of blowfly 37
–, phytin, utilization in fish 35
–, requirement for antlers 70
– – for egg production 60
– – for organic structure 7, 10
– –, of catfish 35
– –, of dog 72
– –, of horses 71, 72
– –, of man 77
– –, of pig 73
– –, of poultry 58–61
– –, of ruminants 66–70
– –, of subhuman primates 74, 75
– –, of turkeys, factors affecting 60, 61
–, sources, availability 60
Photosynthesis, carbon requirement for 8
–, role of potassium in 16
Phthalthrin, toxicity 162
Phyletic review of primitive groups in equilibrium with sea water 29
Phylogenetic review of toxicants, acoelomates 168
– – –, diploblasts 168
– – –, entercoela 172, 173
– – –, protozoa 165–167
– – –, pseudocoelomates 169, 170
– – –, schizocoela 170–172
Physeter, food habits of 33
Phytase, hydrolysis of phytic acid during baking by 43
Phytate, calcium, availability of 58
–, chelation to minerals 152
–, content, of wheat flour, effect of extraction rate on 43
–, effect on calcium absorption 41
– – on iron absorption 53
– – on phosphorus requirement 35
– – on zinc absorption 120
– – – requirement 101
–, excess, adaptation to 41, 43
– –, disturbances in zinc and calcium metabolism due to 44
– – inhibition of utilization of minerals by 41
–, in flour, hydrolysis by phatase during baking 43
–, phosphorus, absorption of 41

– –, availability in pigs 74
Phytolacca dodecandra 171
Pig, calcium requirement of 44
–, copper deficiency in 106
– – requirement of 101
–, gestation, manganese requirement for 100
–, growth of, effect of copper excess on 148
– – – of manganese toxicity on 114
– –, manganese requirement for 102
– –, zinc requirement for 101
–, magnesium requirement of 45
–, mineral constituents of milk 64
– – requirements of 73, 74
–, reproduction, manganese requirement for 109
– –, zinc requirement for 100
–, selenium deficiency in 24, 110
–, tolerance to iodine excess 114
–, zinc deficiency in 23
– – toxicity in 116
Pigments, respiratory 51, 52
Piperazine, action in nematodes 166, 167
Pizonyx, food habits of 34
Plankton, as source of food for marine birds 33
–, role in nitrogen and phosphorus cycles in the sea 34
Plants, requirement for cobalt 21
– – for copper 22
– – for iron 21
– – for selenium 23
– – for zinc 23
Plastocyanin electron carriers 22
Platyhelminthes, biocides for 166
Pododermatitis, in cattle, effect of zinc on 121
Pogonophores, hemoglobin in 51
Polychaetes, chlorocruorin in 52
–, hemoerythrin in 52
–, iron content of 30
–, regulation of ions in 29
Polycythemia, induced by cobalt 114
Polyuronides, silicon in 11
Pony, mineral requirements of 71, 72
Porifera, metabolism 168
–, toxicity of 168
Porphyria, loss of urinary excretion in 120
Porphyrin, in chlorocruorin 52
Porpoise, as source of food for marine mammals 33
–, food habits of 33
Potassium, catalytic function of 14, 16
–, concentration in decapod crustaceans 29
– deficiency, in children 48
–, excretion in decapod crustaceans 29
–, in serum, effect of magnesium deficiency on 75
–, in urine, of cephalopods 29
–, regulation in echinoderms 29
– – in fishes 30
– – in marine animals 28
– – in mullascs 29
– – in tunicates 30
–, requirement for maintenance of osmotic and electrolyte balance 6
– –, of horses 71, 72
Poultry, calcium requirement of 44
–, gestation and reproduction, manganese requirement for 100
–, growth, calcium and phosphorus requirements for 59, 60
– – – – – for, factors affecting 59, 60
– –, iodine requirements for 102
– –, magnesium requirement for 61
– –, zinc requirement for 102
–, mineral requirements for, factors influencing 58–61
–, reproduction, manganese requirement for 104
–, requirements for calcium, phosphorus, magnesium and iron, table 59
–, selenium deficiency in 110
–, zinc toxicity in 116
Prawns, absorption of minerals in 35
Pre-phenol oxidase, in insects, copper content 96
Pregnancy, calcium deposition in mammals during 65
–, chromium deficiency in 105
–, in ruminants, requirement of major mineral elements for 66

–, iron losses in 76
Pripapuloids, hemoerythrin in 52
2-Propenyl isothiocyanate 50
Propoxure, toxicity 162
Prostate gland, zinc in 92
Protein A 24
–, effect on phosphorus and calcium metabolism 42
– synthesis, need for magnesium in 45
Protein-calorie malnutrition, effect on trace element content of liver 107
Proteus morgani 153
– *vulgaris* 153
Prothrombin synthesis, requirement for manganese 21
Protosphere 8
Protozoa, biocides for 166
–, calcium in 13
–, magnesium requirement of 18
–, metabolism in 165
–, potassium requirement of 16
–, silicon requirement of 11
–, toxicology of 165–167
–, trace element requirements of 93, 94
Protozoan diseases, effectiveness of antibacterial drugs in 165
Pseudocoelomates, toxicology of 169, 170
Pseudomonas, calcium deficiency in 17
Pseudosarcophaga affinis 16
Pteriomorphia 32
Purines, antimetabolites of 153, 154
–, oxidation to uric acid, in worm 97
–, requirement for 9
Pyramethamine, action in protozoa 166
Pyrethrum, action in arthropods 167
Pyridoxine, toxicity 151
Pyrimidines, antimetabolites of 153, 154
–, requirement for 9
Pyruvate carboxylase, activity, effect of manganese deficiency on 20
– –, manganese in 20
Pyruvic kinase, potassium requirement for 16

Quail, zinc deficiency in 113
Quinacrine, action in protozoa 166
Quinic acid, dehydrogenation to benzoic acid 173

Rabbit, calcium absorption in 42
–, growth, manganese requirement for 102
–, iodine toxicity in 114
–, manganese deficiency in 109
–, mineral constituents of milk 64
–, phosphorus absorption in 42
Radial diffusion 43
Radionuclides, in food 158
Ram, testicular function, zinc requirement for 104
Rana, enzymes in 99
Rat, calcium absorption in 42
–, chromium deficiency 105
–, dental caries, prevention by fluorine 110
–, effect of copper excess in 113
–, essentiality of nickel in 22
–, gestation, copper requirement for 99
– –, zinc requirement for 100
–, growth, effect of zinc deficiency on 112
– –, iodine requirement for 102
– –, manganese requirement for 102
– –, silicon requirement for 11, 103
– –, stimulation by chromium 114
– –, tin requirement for 103
– –, vanadium requirement for 103
– –, zinc requirement for 101
–, iron absorption in 55
– – deficiency anemia in 57
–, magnesium deficiency in 46, 47
–, manganese absorption in 109, 119
–, mineral requirements of 72
–, molybdenum toxicity, abolishment by iodine 109
–, phosphorus absorption in 42
–, reproduction, zinc requirement for 103, 104
–, selenium deficiency in 111
– – toxicity in 116
–, susceptibility to norbormade 173
–, tin toxicity in 116
–, tolerance to iodine excess 114
–, trace element requirements of 97
–, vanadium deficiency in 19

– – toxicity in 116
–, zinc toxicity in 116
Recommended daily allowance 148
– – intake of nutrients for UK 76, 77
– dietary allowances for man 75, 76
Recondite toxicity 159
– – of metals 148
Reindeer, ash content of milk 64
Reproduction, effect of zinc deficiency on 112
–, impairment of, in manganese deficiency 109
–, manganese requirement for 100, 104
–, of insects, effect of trace element deficiency on 95
–, of mice, in fluorine deficiency 110
–, of poultry, manganese requirements for 104
–, role of zinc in 23
–, toxicants affecting 159
–, trace element requirements for 103, 104
Reptiles, trace element requirements of 98, 99
–, calcitonin in 40
Resorption rate 43
Respiration, toxicants affecting 159
Respiratory pigments, chlorocruorin 52
– –, hemocyanin 52
– –, hemoerythrin 52
– –, hemoglobin 51, 52
– –, hemovanadin 52
Reticulocytes, transfer of iron to 54
Rhesus monkey 118
Rhizobium, calcium deficiency in 17
Rhodanase, conversion of cyanide and thiosulfate to thiocyanate by 50
Ribonucleic acid, stabilization by nickel 22
Ribosomes, structure, effect of nickel on 22
Rodenticides 172
–, toxicity 167
Rotenone, action in arthropods 167
–, toxicity 173
Rumen microorganisms, effect of zinc on 117
Ruminants, cobalt deficiency in 104, 105
–, growth, cobalt requirement for 102
– –, selenium requirement for 103
–, interaction between molybdenum, copper and inorganic sulfate in 115
–, requirements for major mineral elements 66–71
–, selenium deficiency in 110
– – metabolism in 120
–, sulfur requirement of 70, 71
–, vitamin B_{12} deficiency in 21

Salivary glands, of blowfly, role of calcium in 36, 37
Salmo trutta (L) 36
Salmon, liver arginase, dependence on manganese 98
–, nutritional requirements of 35
Salpa, ionic regulation in 30
Sarcoplasmic reticulum 38
Schistosomiasis, treatment by antimonials 164
Schizocoela, toxicology of 170, 171
Schizophrenia, zinc excretion in 121
Sea anemone, regulation of calcium in 29
– otter, food habits of 33, 34
– urchins, as source of food for marine mammals 34
– water, content of calcium 33
– –, ionic composition 55
– –, ionic content 30
– –, marine animals in equilibrium with 29
Seal, food habits of 33
Selenium, absorption 120
–, as component of glutathione peroxidase 120
–, as essential element 2
–, catalytic function 23, 24
– deficiency 110, 111
– –, alleviation by arsenic 116
– –, growth retardation in 102
–, essentiality of 97
–, function in relation to vitamin E 120
–, interaction with copper 111
–, occurrence in animal tissues 91, 92
–, requiremnt, of mammals 97
– –, of poultry 98

–, relationship to vitamin E 103
– toxicity 115, 116, 151
– –, in grazing animals 24
Seminal tubules, degeneration, in manganese deficiency 109
Senesence, bone resorption in 41
Shearwaters, food habits of 33
Sheep, calcium deposition during pregnancy 65
–, cobalt deficiency in 104, 105
–, copper poisoning in 23, 113
–, effect of cobalt excess in 114
–, growth, effect of selenium deficiency on 103
– –, mineral requirements for 67
–, mineral constituents of milk 64
–, pregnancy, mineral requirements for 69
–, pre-natal development, copper requirement for 99
–, selenium deficiency in 103, 110, 111
–, silicon metabolism in 119
–, zinc deficiency in 112
Shell mineralogy, genetic basis 32
– structure, of bivalves 32
Shells, animal, formation 31
–, of bivalves, mineral content 32
Silicon, absorption 119
–, catalytic function of 15
–, growth-promoting properties of 97
–, occurrence in animal tissues 92
–, requirement for 11
– –, growth 103
– –, in chick 98
Sipunculoids, hemoerythrin in 52
Skeleton, deformities, in manganese deficiency 108
–, of marine animals, major minerals in 30–32
Skin lesions, in magnesium deficiency 46
– –, in zinc deficiency 23, 112
–, loss of iron from 76
–, zinc in 92
Snails, biocides for 167
–, zinc requirements of 23
Soap formation, effect of fat or calcium excess on 151
Sodium, catalytic function of 14–16
–, concentration in decapod crustaceans 29
–, effect on cell permeability 152
– – on growth of crickets 146
–, excretion in decapod crustaceans 29
– fluoracetate, action in rodents 167
–, in urine, of cephalopods 30
– iron pyrophosphate, as source of iron 56
–, regulation in fish 30
– – in marine animals 28
– – in tunicates 30
–, requirement for maintenance of osmotic and electrolyte balance 6
– –, of primitive and zoophagous insects 6
Sow, requirements for calcium 74
Soya protein, effect on iron absorption 75
Soya bean protein, effect on availability of iron 61
Space travel, effect on bone density 74
Sperm, mobility, in selenium deficiency 110
– whale, food habits of 33
Spermatogenesis, effect of manganese deficiency on 109
– – of zinc deficiency on 112
–, zinc requirement for 104
Spider monkey, food habits of 74
Squid, as source of food for marine animals 33
Squirrel monkey, selenium deficiency in 110
Staphylococcus epidermidis 121
Staphylomycin, effect on vinegar eel 170
Starvation, loss of urinary zinc in 120
Sterility, in molybdenum toxicity 115
Stibophen, action in platyhelminthes 166
–, inhibition of phosphofructokinase by 161
Stilbamidine, action in protozoa 166
Stilbazium iodide, action in nematodes 166
Stomach, contents, of trout 36
Streptococcus spp. 121
Stress, due to, trace element deficiency 94
–, zinc and 121
Stromateus niger 36
Strontium, in *Acantharia* 13

Strychnine, action in rodents 167
–, toxicity of 173
Sturgeon fry, trace element dynamics in 98
Subhuman primates, requirements for major minerals 74, 75
Succinic dehydrogenase, requirement for iron 21
Sulfanilamide, and *p*-aminobenzoic acid 152
Sulfate, absorption, in marine animals 32
– –, in trout 35
–, catalytic function of 14
–, concentration in decapod crustaceans 29
–, content of sea water 33
–, effect on copper absorption 117
– – – requirement 101
– – on molybdenum toxicity 24
–, excretion in decapod crustaceans 29
– – in urine 48
–, in urine, of cephalopods 29
–, inorganic, effect on molybdenum absorption 119
–, interrelationship with copper and molybdenum 23
–, protective action against molybdenum toxicity 115
–, regulation, in fishes 30
– – in marine animals 28
– – in primitive marine animals 29
– – in tunicates 30
–, sulfur, metabolism 51
–, utilization by mammals 70
Sulfated compounds, metabolism of 49
– polysaccharides, sulfur in 48
Sulfide sulfur, metabolism of 51
Sulfite oxidase, activation by molybdenum 97
– –, molybdenum content 119
Sulfolipids, sulfur in 48
Sulfur amino acids, relation to selenium 24
– – –, synthesis by microorganisms in rumen 70
–, catalytic function 19
– deficiency, in ruminants 70
–, in goitrogenic compounds 50
–, in ruminant nutrition 70, 71
–, inorganic, content of foods 48
–, metabolism 48–51
–, requirement for organic structure 7, 10
– – for synthesis of chondroitin sulfates 13
– –, of horses 71, 72
Superoxide dismutase, presence of copper in 22
Supportive structures 10
Suramin, action in nematodes 166
Surgical trauma, loss of urinary zinc in 120
Swallow-tailed gulls, food habits of 33
Symbiosis, microbial, vitamin B_{12} synthesis due to 21
Synthetic nutrients, unconventional 9

Tanok, phytate content of 44
Taste, effect of zinc on 121
Taurine, conversion of sulfate sulfur to 51
–, excretion in urine 48
–, sulfur in 48
TDE, resistance of insects to 173
Teeth, calcified tissues in 38
–, fluorine in 91
–, of higher animals, calcium and phosphorus in 37, 38
–, radular, of chitons, iron in 30
–, zinc in 92
Teleosts, insecticide toxicity in 162
–, ionic regulation in 30
Temperature, effect on nutrient requirements of microorganisms 94
– – on serum magnesium 46
Tenebrio molitor 95
Teratologic changes, in embryo, in selenium toxicity 115
Terebella, hemoglobin in 52
Testes, function, zinc requirement for 104
Testicular atrophy, in zinc deficiency 112
– defects, in selenium deficiency 111
Tetany, due to magnesium deficiency 46
Tetrachloroethylene, action in nematodes 166
Tetrahymena geleii 16, 18
Tetrahymena, inorganic requirements of 94
Tetramisole, action in nematodes 166

Tetramizole, selective toxicity against round worms 169
Theobalda incidens 16
Thiabendazole, action in nematodes 166
Thiamine, occurrence and role 50
–, requirement for sulfur 19
–, sulfur in 10, 48
Thiocyanate, as iodine antagonist 102
–, production from cyanide 50
–, sulfur in 48
Thiolhistidine, sulfur in 48
Thiosulfate, excretion in urine 48
–, oxidation of hydrogen sulfide to 51
Thiouracil, effect on iodine utilization 50
Thrombosis, due to magnesium deficiency 46
Thyroid gland, calcitonin in 40
– –, effect of molybdenum and iodine on 109
– –, iodine in 91
– hormone, function, effect of iodine deficiency on 109
– – –, in amphibia 99
– –, synthesis, role of copper and manganese 109
Thyroxine, effect on synthesis of chondroitin sulfates 49
–, synthesis, requirement for iodine 18
Tilapia, artificial feeding of 35
Tin, absorption 119
–, as essential element 2
–, catalytic function 24
–, growth-promoting properties of 97
–, requirement for growth 103
Tongue, lesions, in zinc deficiency 112
Toothed whales, food habits of 33
Toxaphen, action in insects 173
–, toxicity 163
Toxicants, adaptation to 173, 174
–, classification 151
– – on basis of lethal doses 150
–, definition 144
–, ingested, activity spectrum 144–178
– –, antimetabolites 152–154
– –, essential nutrients 151–152
– –, food toxicants 154–158
– –, other toxicants 158–160
–, primary 159, 160
–, secondary 159, 160
–, stimulatory properties of 148
Toxicity, of excess of nutrients 148
–, of trace elements 104–117
–, selective 160
Trace elements 13, 87–143
– –, cationic, excretion of 2
– –, deficiency of, in poultry 98
– –, effects of deficiency and toxicity 104–117
– –, functional requirements 99–104
– –, interactions, in insects 95
– –, metabolism and biological role 117–122
– –, nomenclature 87–89
– –, occurrence 89–93
– –, qualitative requirements, invertebrates 93–97
– – – –, vertebrates 97–99
– –, toxicity 113–117
Transferase, role of manganese in 20
Transferrin, content of blood plasma 53
–, iron content of 51
–, role in iron absorption 53
Transitional elements 2
– –, difference from calcium 31
Translocation, in plants, role of potassium in 16
Travisia, hemoglobin in 52
Trehalose, content, in insects 5
Trematode, inhibition of phosphofructokinase in 161
Tribolium confusum 95, 96
Trichlorophenal, action in nematodes 167
Trichomonicides, action on flagellates 167
Tridaena gigas 32
3-Trifluoromethyl-4-nitrophenol 172
Triglyceridemia, in anemia 57
Triiodothyronine synthesis, requirement for iodine 18
Trimethylamine, content of plasma, in elasmobranchs 5
Tropical birds, food habits of 33
Trout, absorption of minerals in 34, 35

–, food habits of 36
Tryptophan desmolase, zinc in 23
– oxygenase, requirement for iron 21
Tubinares 33
Tumor development, effect of zinc on 113
Tuna, ovaries, trace elements in 98
Tungsten, effect on molybdenum requirement 98
Tunicates, cellulose in 10
–, ionic regulation in 30
Turbatrix aceti 169, 170
Turkey, egg production, manganese requirement for 100
–, growth, magnesium requirement for 61
– –, selenium requirement, effect of vitamin E on 103
–, growth-promoting properties, of molybdenum in 98
–, reproduction, manganese requirement for 104
–, requirement for calcium and phosphorus 60, 61
– – for major mineral elements, table 59
–, zinc deficiency in 113
Turtle, blood serum, magnesium content 45

Ultimobranchial cells, regulation of blood calcium by 40
– glands, calcitonin in 40
Unconventional sources of food 9
United Kingdom, Department of Health and Social Security 76, 77
– –, National Food Survey 76
– –, recommended daily intakes of major minerals 77
United States, National Academy of Sciences 75
– – – Research Council 76, 77
– –, recommended daily dietary allowances for major minerals 77
Urea, as source of nonspecific nitrogen 9
–, content of plasma, in elasmobranchs 5
Uric acid, from purines, in worms 97
Uricase, association of copper with 118
–, in amphibia 99
–, lack of, in worms 97
Urinary tract, losses of iron from 76
Urine, of cephalopods, differential ion excretion in 29
–, of decapod crustaceans, ion excretion in 29

Valine, competition with leucine and isoleucine 152
Vanadium, absorption 119
–, as essential element 2
–, catalytic function 19
–, chromogen 52
– deficiency 19
–, growth-promoting properties of 97
–, occurrence in animal tissues 92
–, requirements for growth 103
– –, of chicks 98
–, toxicity 116
Vascularisation, abnormal, in selenium deficiency 112
Vasodilation, in magnesium deficiency 46
Vertebrates, toxicology of 172
–, trace element requirements of 97–99
Vinegar eel, effects of antibiotics on 169
– – – of stahpylomycin upon survival to heat stress 170
1,5-Vinyl-2-thiooxazolidone 50
Viprynium, action in nematodes 166
Vitamin A deficiency, impairment of synthesis of chondroitin sulfate in 49
– –, toxicity, in laboratory animals 151
– –, utilization and storage, role of zinc in 121
– B_1, *see* Thiamin
– B_{12}, coenzymes 21
– –, depletion of stores in ruminants, in cobalt deficeincy 105
– –, requirement, of amphibia 99
– –, synthesis, in microorganisms 21
– – – in ruminants, cobalt requirement for 97
– B_{12}-containing enzymes, requirement for cobalt 21
– C deficiency, impairment of synthesis of chondroitin sulfate in 49
– –, toxicity, in man 151

– – –, in rats 151
– D deficiency, changes in mitochondria in 39
– –, effect on availability of phytate phosphorus 43
– – – on calcium requirement of pig 74
– – – on phytin phosphorus absorption 41
– – – on requirements for calcium and phosphorus 44, 58
– –, interdependence of parathyroid hormone and 39
– –, metabolism, disorders in, osteoporosis due to 76
– –, toxicity in infants 151
– E, alleviation of selenium deficiency symptoms by 112
– –, relationship to selenium 24, 103
– –, selenium function in relation to 120
Vitamins, antimetabolites of 153, 154
–, fat-soluble, absorption of, effect of phosphates on 151
–, requirement for 9

Wallago attu 36
Walruses, as source of food for marine mammals 33
–, food habits of 33
Water balance, in animals 6
–, establishment upon earth 7
–, metabolic 6
–, pollutants, toxicity of 165
–, toxicity of 151
Weight reduction, in obese subjects, magnesium deficiency in 48
Whales, as source of food for marine mammals 33
–, food habits of 33
Wheat fluor, iron enrichment of 55
White Amur roe, cobalt toxicity in 114
– muscle disease 111
– – –, in selenium deficiency 110
Wool growth, effect of zinc deficiency on 112
Worms, myoglobin in 52
–, trace element requirements of 96
Wound healing, effect of anemia on 57
– – – of zinc deficiency on 112
– –, role of chondroitin sulfates in 49
– – – of zinc in 23

Xanthine oxidase, absence in protozoa 94
– –, activation by molybdenum 97
– –, in amphibia 99
– –, molybdenum content 24, 119
– –, requirement for iron 21
– –, sulfur in 49
Xenopus, enzymes in 99

Yellow tail, artificial feeding of 35

Zero equivalent point 149
Zinc, absorption 120
– –, effect of phosphate on 151
– –, in pig, effect of minerals on 74
– – – – of phytate on 74
–, as essential element 2
–, availability, effect of calcium on 104
–, catalytic function 23
–, competition with other minerals 150
–, content, of blood, effect of zinc deficiency on 112
– –, of liver, effect of nickel excess on 115
– – –, in protein-calorie malnutrition 107
– deficiency 23, 112, 113
– –, effect on polynucleotide metabolism 122
– –, in *Euglena gracilis* 94
– –, in poultry 98
–, effect on calcium storage 33
– – on copper absorption 117
–, excess, interference with utilization of copper 58
–, interaction with copper 95
–, interrelationship with iron, calcium and copper 22
–, metabolic stores 101
–, metabolism, disturbances, due to high phytate 44
–, occurrence in animal tissues 92, 93
–, protection against copper poisoning by 113

Index vol. 1

BIRD, H. R.: Antibiotics *1* 168
BIRD, H. R.: Hormones *1* 181
BISHOP, D. G.: Lipids and Lipid Metabolism *1* 74
COATES, M. E.: The Water-Soluble Vitamins and Other Accessory Food Factors *1* 136
KRONFELD, D. S. and VAN SOEST, P. J.: Carbohydrate Nutrition *1* 23
LUCKEY, T. D.: Introduction to Comparative Animal Nutrition *1* 1
THOMPSON, J. N.: Fat-Soluble Vitamins *1* 99

–, requirement for gestation 100
– – for reproduction 104
– –, of amphibia 99
– –, of fish 98
– –, of insects 95, 96
– – –, effect of cadmium on 95
– –, of mammals 97
– –, of protozoa 94
– –, of worms 97
–, toxicity 116, 117
–, utilization of, inhibition by phosphate 41
Zooplankton, as source of food for trout 36